Inorganic
Coordination Compound

D0881911

Nobel Prize Topics in Chemistry

A Series of Historical Monographs on
Fundamentals of Chemistry

Editor
Johannes W. van Spronsen
(The Hague and University of Utrecht)

Advisory Board
G. Dijkstra (Utrecht)   N. A. Figurowsky (Moscow)
F. Greenaway (London)   A. J. Ihde (Madison)
E. Rancke-Madsen (Copenhagen)   M. Sadoun-Goupil (Paris)
Irene Strube (Leipzig)   F. Szabadvary (Budapest)
T.J. Trenn (Munich)

Nobel Prize Topics in Chemistry traces the scientific development
of each subject for which a Nobel Prize was awarded in the light of
the historical, social and political background surrounding its
reception. In every volume one of the Laureate's most significant
publications is reproduced and discussed in the context of his life
and work, and the history of science in general.

This major series captures the intellectual fascination of a field
that is too often considered the domain of specialists, but which
nevertheless remains a significant area of study for all those
interested in the evolution of chemistry.

Current titles
Stereochemistry   *O. Bertrand Ramsay*
Transmutation: natural and artificial   *Thaddeus J. Trenn*
Inorganic Coordination Compounds   *George B. Kauffman*

Inorganic Coordination Compounds

George B. Kauffman

California State University, Fresno, California, USA

HEYDEN

LONDON · PHILADELPHIA · RHEINE

Heyden & Son Ltd, Spectrum House, Hillview Gardens, London NW4 2JQ, UK
Heyden & Son Inc., 247 South 41st Street, Philadelphia, PA 19104, USA
Heyden & Son GmbH, Devesburgstrasse 6, 4440 Rheine, West Germany

---

British Library Cataloguing in Publication Data

Kauffman, George B.
    Inorganic coordination compounds – (Nobel prize topics in chemistry)
    1. Chemistry, Inorganic
    I. Title. II. Series
    546   QD151.2   80-42031

    ISBN 0-85501-683-3
    ISBN 0-85501-684-1 Pbk

Printed in Great Britain by Cambridge University Press, Cambridge

Contents

List of plates

Foreword

By Sir Geoffrey Wilkinson, FRS
(Nobel Laureate in Chemistry, 1973)

In January 1954 I arrived in Copenhagen to spend a semester's leave from Harvard in Professor Jannik Bjerrum's laboratory in the old Polyteknisk Laereanstalt on Sølvgade. I was given a rather old-fashioned room with a fume cupboard whose draft was provided by a Bunsen burner. This room I later found out had been Sophus Mads Jørgensen's, and it had survived virtually unchanged, along with hundreds of his samples. Shortly afterwards, I read in my father-in-law's library the delightful account by John Read in *Humour and Humanism in Chemistry* of the time he spent in Alfred Werner's laboratory in Zürich. For the first time I began to take an interest in the history of inorganic chemistry because of this almost accidental acquaintance with the two giants of complex compounds – Jørgensen, the supreme master of synthesis, and Werner, the man with the right ideas of structure.

In the present book, Professor Kauffman, winner of the 1978 Dexter International Award in the History of Chemistry, gives us not only the story of the celebrated Werner–Jørgensen dialogue but also accounts of the struggles of many of the earlier workers in the field, some of whom I have to admit I had not known previously. His account makes me, at any rate, wish to know more of the life and times of these pioneers in what is surely one of the most important fields of chemistry.

The position of Werner is unique. Although his views on the nature of the chemical bond were not correct, his epoch-making concept of the structure of complex compounds, which brought order to chaos, has had a profound effect not only on inorganic chemistry, but on the whole of chemistry, including much of the life sciences.

Those who have humbly attempted to follow in Werner's footsteps can only feel grateful to the scholarship of Professor Kauffman for his

fascinating and sympathetic account of our scientific forebears. It should
be compulsory reading for all students.

Geoffry Wilkinson

Professor of Inorganic Chemistry
Imperial College of Science and
Technology, London

General Editor's Preface

Nobel Prizes in Chemistry have been awarded almost every year since 1901, and the topics covered by these awards have touched upon almost every subject in chemistry. 'Nobel Prize Topics in Chemistry' plans to cover the history of each subject for which a Nobel Prize was awarded and to place particular emphasis on the life and work of the Nobel Prize winner himself. In this way the planned Series will come to describe the whole history of chemistry. The concept is to take one of each Nobel laureate's most significant publications, to reprint it (as an English translation if appropriate), discuss it, and then to place it within the context of the laureate's life and works in particular and the history of science in general, if possible going as far back as Egyptian, Babylonian and Greek antiquity. The Series will also look at possible future developments.

Contributions are presented in such a manner that a non-specialist background will suffice for the text to be comprehensible. The intention is to make as many readers as possible aware of and conversant with the problems underlying the development of various areas in the field of chemistry. Each volume attempts to give a smooth outline of the particular topic under consideration, uninterrupted by continual footnotes and references in the text.

The Series is not, in the first instance, aimed at the professional historian, but rather at the chemist, the research worker and the non-specialist who wishes to bring himself up to date on the historical background of one or more areas of chemistry. The student of chemistry and the historian or sociologist who for research wishes to focus broadly on one of the most spectacular disciplines in the natural sciences can obtain a wide-ranging historical knowledge — knowledge which forms part of the general history of mankind and which can also be used to examine the reciprocal relationship between chemistry and society as a whole. The authors are well-known historians of chemistry or chemists with a solid knowledge of history. Among the latter group, occasionally a Nobel Prize winner will be our author. Nobel Prize winners are also contributing forewords to many of our volumes.

For certain parts of the texts a reasonable knowledge of chemistry – in some cases, the reading of formulas, for example – is required, but the needs of the non-chemist have been anticipated by including in each volume a glossary through which the reader can revise or extend his chemical knowledge. Each volume also contains a chronology of significant events and a detailed bibliography.

It is the aim of the Editor that the readers of this series should obtain a clear idea of the particular experiences of a chemist in performing his research – research which sometimes led to a discovery of the greatest significance for humanity. It is thus intended not only to focus exclusively here on the main historical-chemical facts but also to understand the chemist as a human being and to look at the circumstances which led to his discoveries. However, to understand the social and eventually the political background of these historical developments, the reader must inform himself of the facts presented here. This series thus aims to capture the intellectual fascination of a field that is too often considered to be the domain of specialists, but which nevertheless remains an area of proven intellectual adventure for all those who consider the quest for understanding the highest point to which man can aspire.

It is our sincere wish that the individual volumes in this series will realize these aims and intentions.

J.W. van SPRONSEN
The Hague

Author's Preface

The field of coordination chemistry encompasses a great diversity of substances and phenomena. If we are to examine its history within the bounds of a short book, we must impose certain limits on our survey. I have therefore chosen to confine myself to the highlights and forego many of the interesting but minor events. Thus this extremely selective treatment will be limited to a discussion of what I consider the major discoveries, both experimental and theoretical, from the beginning of the nineteenth century to the first third of the twentieth century. In view of the central role of Alfred Werner, the founder of coordination chemistry, we will naturally devote considerable time to his monumental achievements, but we will also look at some of the achievements of lesser known contributors.

There is, of course, a disadvantage other than lack of completeness in the topical approach that I have chosen. Such a selection of highlights may tend to give you the false but alluring impression of the progress of science as a smooth, unbroken, steadily ascending line. On the contrary, the history of science, in common with all history, is not a continuous function but an erratic one that proceeds in spurts with more dead ends or even backward steps than we generally care to admit. With these warnings in mind, I hope that this book will take you on an exciting journey that will include a mixture of chemistry, biography and history which should enrich your background in what today is one of the most challenging and active branches of inorganic chemistry.

My task in writing this book was simplified by my ability to draw upon my previously published articles and books on Werner and coordination chemistry. (These are listed separately in Appendix D.) It is my pleasure to acknowledge the generous financial assistance of both the History and Philosophy of Science Program of the National Science Foundation Division of Social Sciences (Grants GS-74 and GS-1580) and of the

American Philosophical Society (Penrose Fund, Grant 3255; Johnson Fund, Grant 876). The first mentioned grants from these two organizations permitted me to spend the 1963–64 academic year at Universität Zürich studying Alfred Werner's life and work.

I also wish to thank the American Chemical Society for permission to quote in Chapter 3 from their translation of Werner's lecture of 1912. I am also pleased to acknowledge the assistance of the John Simon Guggenheim Memorial Foundation in the form of a Guggenheim Fellowship and the California State University, Fresno in the form of a sabbatical leave. I am indebted to Robin D. Myers, my research assistant, for reading the manuscript and to the California State University, Fresno Research Committee for typing assistance. Mrs Donna Hamm, Mrs Elsie Taylor and Mrs Joan Jeffries deserve my gratitude for typing the manuscript. I am also indebted to Professor John C. Bailar Jr for assistance in compiling Appendix A. The photographs in this volume are the work of Robert Michelotti, whose fine technical assistance I have called upon through the years. Last but not least, to my wife Laurie go my thanks for inspiration, understanding and encouragement.

GEORGE B. KAUFFMAN
California State University, Fresno,
Fresno, California 93740,
June 1981   USA

Acknowledgements

Thanks are due to the following for permission to reproduce copyright material: Dover Publications, Inc., for general extracts from the present author's *Classics in Coordination Chemistry, Part II* (1976) and *Part III* (1978); the Société Chimique de France and the American Chemical Society for the English translation of 'Sur les composés métalliques à dissymétrie moléculaire', from *American Chemical Journal* 48, 314–336 (1912); Springer-Verlag for Plates 1 and 2, from the present author's *Alfred Werner, Founder of Coordination Chemistry* (1966); Verlag Chemie GmbH for Plate 4, from W. Prandtl, *Deutsche Chemiker in der ersten Hälfte des neunzehnten Jahrhunderts* (1956); Macmillan Press Ltd for Plate 5, from J. R. Partington, *A History of Chemistry*, Vol. 4 (1964); the US National Academy of Sciences for Plates 7 and 8, from the *National Academy of Sciences Biographical Memoirs*, Vol. 4 (1902) and Vol. 7 (1910); the Division of Chemical Education, American Chemical Society, for Plate 9, from *J. Chem. Educ.* 11, 281 (1934); Battenberg Verlag München for Plates 10 and 12, from R. Sachtleben and A. Hermann, *Von der Alchemie zur Grossynthese: Grosse Chemiker* (1960); the National Historic Museum, Frederiksborg, Denmark, for Plate 15; Bassano Studios, London, for Plate 25, from *Proceedings of the Chemical Society* 312 (1958); Argent Studios, London, for Plate 27, from *Biographical Memoir of Fellows of the Royal Society*, Vol. 6 (1960).

For

Laurie

My darling wife and lifelong companion,
with equal parts of love and lust

'Die Geschichte der Wissenschaft ist die Wissenschaft selbst.'

Goethe

1

Definitions of fundamental terms and concepts

Coordination chemistry is, quite simply, the chemistry of coordination compounds. And what is a coordination compound? Before proceeding any further, we had better define our terms (Kauffman, 1977c[†]). Some typical coordination compounds are shown in Table I. A *coordination compound* may be defined as a compound containing a central atom or ion to which are attached molecules or ions whose number usually exceeds the number corresponding to the oxidation number or valence of the central atom or ion. As you can see from the examples, the *central atom* or center of coordination is usually a transition metal, that is, an element from one of the subgroups or B groups of the periodic table. The coordinated groups are called *ligands*. They may be neutral molecules*, such as ammonia in compounds 1, 4, 5 or 8, ethylenediamine as in compounds 2 and 7, or water as in compound 10. They may also be *ions** such as chloride in compounds 5, 6 and 8, bromide as in compound 2, oxalate as in compound 3, or cyanide as in compound 9. *Metal–ammines*, in which ammonia molecules are coordinated to a central metal ion, are among the commonest coordination compounds.

Ligands are attached to the central atom by means of what are called *coordinate bonds* or *coordinate covalent bonds*. This type of bond was first postulated by the American chemist Gilbert Newton Lewis in 1923. A coordinate bond differs from an ordinary covalent bond as follows.

---

* Throughout the text, terms designated with an asterisk are defined in the Glossary (pp. 178–181).

[†] For literature see Appendices C and D.

Inorganic Coordination Compounds

TABLE I

Typical coordination compounds

Compound[a] Systematic IUPAC name	Oxid. no.	Coord. no.	Ionic charge
(1) $[Co(NH_3)_6]Cl_3$ Hexaamminecobalt(III) chloride	3+	6	3+
(2) $[Co(\underline{H_2NCH_2CH_2NH_2})_2Br_2]Cl$ Dibromobis(ethylenediamine)cobalt(III) chloride	3+	6	1+
(3) $K_3[Co(\underline{C_2O_4})_3]$ Potassium trioxalatocobaltate(III)	3+	6	3−
(4) $[Pt(NH_3)_4](NO_3)_2$ Tetraammineplatinum(II) nitrate	2+	4	2+
(5) cis- or trans- $[Pt(NH_3)_2Cl_2]$ Dichlorodiammineplatinum(II)	2+	4	0
(6) $K_2[PtCl_4]$ Potassium tetrachloroplatinate(II)	2+	4	2−
(7) $[Pt(\underline{H_2NCH_2CH_2NH_2})_3]Cl_4$ Tris(ethylenediamine)platinum(IV) chloride	4+	6	4+
(8) cis- or trans- $[Pt(NH_3)_2Cl_4]$ Tetrachlorodiammineplatinum(IV)	4+	6	0
(9) $K_4[Fe(CN)_6]$ Potassium hexacyanoferrate(II)	2+	6	4−
(10) $[Cu(H_2O)_4]SO_4 \cdot H_2O$ Tetraaquacopper(II) sulfate monohydrate	2+	4	2+

[a] Bidentate or chelate ligands are underlined.

$$H \overset{\times\times}{\underset{\times\times}{\overset{\times}{\underset{\times}{Cl}}}}$$

$$\left[\begin{array}{c} H_3N: \diagdown \underset{Cu}{\diagup} :NH_3 \\ H_3N: \diagup \diagdown :NH_3 \end{array} \right]^{2+}$$

Covalent bond   Coordinate bonds

One electron donated from each atom   N atom = Donor or
                                                         Lewis base
x = electrons from chlorine atom Cu^{2+} ion = Acceptor or
                                                         Lewis acid
• = electron from hydrogen atom

Plate 1. Alfred Werner (1866–1919) [Kauffman, 1966c, frontispiece]

In an ordinary covalent bond each of the bonded atoms contributes one electron to the electron pair that forms the bond. In the coordinate bond, on the other hand, the coordinating atom or ligand, here called the *donor*, donates a pair of electrons to the central atom, here called the *acceptor*. The bond is often depicted by an arrow proceeding from the donor atom to the acceptor atom. Donor atoms are usually nonmetals, the most common being nitrogen, oxygen and sulfur. The difference between a coordinate bond and an ordinary covalent bond consists solely in its mode of formation. As even Alfred Werner, the founder of coordination chemistry (Plate 1), recognized, once the bond is formed, the two types are identical.

The entire aggregate of central atom and ligands is sometimes called a *complex*. Ligands may be *unidentate*, that is, they may possess only one coordinating atom. The ammonia molecule or the chloride ion are examples of unidentate ligands. Ligands may also be *bidentate* or *chelate*, that is, they may possess two coordinating atoms. In Table I the ethylenediamine molecule as shown in compounds 2 and 7 and the oxalate ion as shown in compound 3 are common chelate groups, which necessarily form a *ring* with the central atom. *Polydentate* or *multidentate* ligands containing more than two coordinating atoms are also possible.

Most complexes are *mononuclear*, that is, they contain only one central atom. All the compounds in Table I are mononuclear. However, *polynuclear* complexes (Kauffman, 1973b), that is, ones with two or more central atoms, are known and one of these (see pp. 133-134) formed the pinnacle of Werner's life's work. Electrically, a complex may be *positive*, as in compounds 1, 2, 4, 7 and 10, *negative*, as in compounds 3, 6 and 9, or even *neutral*, as in compounds 5 and 8. The charge on the complex merely depends on the balance between the charges of the central atom and of the ligands. Although partial dissociation may occur in some cases, the complex usually tends to remain as a discrete unit, even in solution. Compounds in which the negative complex contains only one type of ligand, such as compounds 3 and 9 in Table I, are known as *double salts**, and compounds containing coordinated water molecules such as compound 10 are known as *metal salt hydrates*. Although double salts and metal salt hydrates were once considered as distinctive types of compounds, Werner demonstrated their close relationship to metal–ammines, and he showed that they should all be regarded as coordination compounds (Werner, 1893).

The total number of ligands surrounding and bonded to a central atom is known as the *coordination number* of the central atom (Werner, 1893). Coordination numbers from one to ten are known, but the most common are six and four. Just as in the field of organic chemistry, ligands are oriented about the central atom in definite spatial *configurations*. For example, complexes with coordination number six usually possess an octahedral configuration, while those with coordination number four are usually square planar or tetrahedral. Because of this orientation in space, the existence of *isomers** — compounds with the same percentage composition and hence the same formula but with different properties — is possible in certain cases. These isomers are of the type known as *stereoisomers*, whose isomerism results from differences in the spatial arrangement of atoms or groups of atoms (configuration). Both types of stereoisomers, viz. *geometric isomers** (Kauffman, 1975a) and *optical isomers** (Kauffman, 1974b), were prepared by Werner to support his theory. Of these, the differences between the latter are the most subtle; each of a pair of asymmetric optical isomers is the exact but nonsuperimposable *mirror image* of the other. The usual laboratory synthesis of an asymmetric substance results in an equimolar mixture of *dextro* (abbreviated *d* or (+) and *levo* (abbreviated *l* or (−)) optically active isomers, known as a *racemic mixture* or *racemate**. The process of separating these isomers (antipodes)* from the mixture is known as *resolution* (see p. 127 *ff*. for details). These concepts will be dealt with in more detail in Chapters 3 and 6. Suffice it to say here that Werner showed that the carbon atom has no monopoly on isomerism. In fact, coordination compounds exhibit many types of stereoisomerism that have no counterparts in organic chemistry.

Also included in Table I are the names of typical coordination compounds according to the rules adopted by the International Union of Pure and Applied Chemistry (IUPAC), which are modified from Werner's original proposals (Werner, 1897). Two other systems of nomenclature are sometimes encountered, especially in the older literature. When naming coordination compounds, early chemists were like the children of Israel, who called the food miraculously supplied to them in the wilderness manna, 'for they wist not what it was'. Similarly, since the true constitutions and configurations of these compounds were unknown until the advent of Werner's coordination theory, complexes were often named after their discoverers (see Table II) or after their colors (see Table III). The color nomenclature was introduced by the Frenchman Edmond Fremy (1852).

According to the IUPAC system, the names of the ligands (modified by di, tri, tetra etc. for simple groups or bis, tris, tetrakis etc. for complex groups to indicate their number) are prefixed to the name of the central metal atom. The names of complex ligands are enclosed in parentheses. The names of negative ligands end in -o and precede the names of neutral ligands, which have no characteristic ending. Coordinated ammonia is designated *ammine*, and coordinated water is designated *aqua*. The oxidation state of the central atom is designated by a Roman numeral enclosed in parentheses. With complex anions, the suffix -*ate* is used, followed by the Roman numeral in parentheses.

Coordination compounds are of great theoretical importance as we shall see in Chapters 2–7, but they are also of great practical utility as well. Coordinating agents are used in metal ion sequestration or removal, solvent extraction, dyeing, leather tanning, electroplating, catalysis, water softening and other industrial processes too numerous to mention here. In fact, new practical applications for them are found almost daily. Vitamin B_{12} is a coordination compound of cobalt, the hemoglobin of our blood is a coordination compound of iron, the hemocyanin of invertebrate animal blood is a coordination compound of copper and the chlorophyll of green plants is a coordination compound of magnesium. Thus coordination compounds are obviously of tremendous significance in biochemistry.

Because of the importance of Werner's research on these compounds, they are sometimes referred to as 'Werner complexes'. We shall now consider Werner's life and work.

TABLE II

Some coordination compounds named after their discoverers

Name	Formula
Anderson's Salt Pyridinium pentachloropyridineplatinate(IV)	$C_5H_5NH[Pt(C_5H_5N)Cl_5]$
Chugaev's Salt Chloropentaammineplatinum(IV) chloride	$[Pt(NH_3)_5Cl]Cl_3$
Cleve's Salt[a] cis-Tetrachlorodiammineplatinum(IV)	$cis\text{-}[Pt(NH_3)_2Cl_4]$
Cleve's Triammine Chlorotriammineplatinum(II) chloride	$[Pt(NH_3)_3Cl]Cl$
Cossa's First Salt Potassium trichloroammineplatinate(II)	$K[Pt(NH_3)Cl_3]$
Cossa's Second Salt Potassium pentachloroammineplatinate(IV)	$K[Pt(NH_3)Cl_5]$
Drechsel's Chloride Hexaammineplatinum(IV) chloride	$[Pt(NH_3)_6]Cl_4$
Durrant's Salt Potassium tetraoxalato- di-μ-hydroxodicobaltate(III) trihydrate	$K_4\left[(C_2O_4)_2Co\begin{smallmatrix}OH\\ \\HO\end{smallmatrix}Co(C_2O_4)_2\right]\cdot 3H_2O$
Erdmann's Salt Ammonium trans-tetranitrodiamminecobaltate(III)	$NH_4\ trans\text{-}[Co(NH_3)_2(NO_2)_4]$
Fischer's Salt Potassium hexanitrocobaltate(III)	$K_3[Co(NO_2)_6]$
Gerhardt's Salt[a] trans-Tetrachlorodiammineplatinum(IV)	$trans\text{-}[Pt(NH_3)_2Cl_4]$
Gibbs' Salt (also called Erdmann's trinitrite) meridional(1, 2, 6)-Trinitrotriamminecobalt(III)	$[Co(NH_3)_3(NO_2)_3]$
Gros' Salt trans-Dichlorotetraammineplatinum(IV) chloride	$trans\text{-}[Pt(NH_3)_4Cl_2]Cl_2$

Name	Formula
Litton's Salt Sodium tetrasulfitoplatinate(II)	$Na_6[Pt(SO_3)_4]$
Magnus' Green Salt[b] Tetraammineplatinum(II) tetrachloroplatinate(II)	$[Pt(NH_3)_4][PtCl_4]$
Magnus' Pink Salt[b] Chlorotriammineplatinum(II) tetrachloroplatinate(II)	$[Pt(NH_3)_3Cl]_2[PtCl_4]$
Morland's Salt Guanidinium *trans*-tetrathiocyanato- diamminechromate(III)	$(NH_2)_2C=NH_2$ *trans*- $[Cr(NH_3)_2(SCN)_4]$
Peyrone's Salt[b, c] *cis*-Dichlorodiammineplatinum(II)	*cis*-$[Pt(NH_3)_2Cl_2]$
Recoura's Sulfate Chloropentaaquachromium(III) sulfate	$[Cr(H_2O)_5Cl]SO_4$
Reinecke's Salt Ammonium *trans*-tetrathio- cyanatodiamminechromate(III)	NH_4 *trans*-$[Cr(NH_3)_2(SCN)_4]$
Reiset's First Chloride Tetraammineplatinum(II) chloride	$[Pt(NH_3)_4]Cl_2$
Reiset's Second Chloride[b, c] *trans*-Dichlorodiammineplatinum(II)	*trans*-$[Pt(NH_3)_2Cl_2]$
Vauquelin's Salt Tetraamminepalladium(II) tetrachloropalladate(II)	$[Pd(NH_3)_4][PdCl_4]$
Zeise's Salt Potassium trichloroethyleneplatinate(II) monohydrate	$K[Pt(C_2H_4)Cl_3]\cdot H_2O$

[a] These compounds are geometric stereoisomers.

[b] These compounds are structural isomers of the type called polymerization isomers by Werner. They have the same empirical formula but have formula weights that are multiples of the same formula weight.

[c] These compounds are geometric stereoisomers.

TABLE III

Some names of complex ions based on color

Color and IUPAC name	Color	Formula
Luteo[a] Hexaamminecobalt(III)	Yellow or orange	$[Co(NH_3)_6]^{3+}$
Purpureo[b] Chloropentaamminecobalt(III)	Purplish red	$[Co(NH_3)_5Cl]^{2+}$
Roseo Aquapentaamminecobalt(III)	Pink or red	$[Co(NH_3)_5H_2O]^{3+}$
Xantho[c] Nitropentaamminecobalt(III)	Brownish yellow	$[Co(NH_3)_5NO_2]^{2+}$
Isoxantho[c] Nitritopentaamminecobalt(III)	Red	$[Co(NH_3)_5ONO]^{2+}$
Tetrammineroseo Diaquatetraamminecobalt(III)	Pink or red	$[Co(NH_3)_4(H_2O)_2]^{3+}$
Violeo[b, d] cis(or 1, 2)-Dichlotetra- amminecobalt(III)	Violet or blue	$cis\text{-}[Co(NH_3)_4Cl_2]^+$
Praseo[b, d] trans(or 1, 6)-Dichloro- tetraamminecobalt(III)	Green	$trans\text{-}[Co(NH_3)_4Cl_2]^+$
Flavo[d] cis(or 1, 2)-Dinitrotetra- amminecobalt(III)	Brownish yellow	$cis\text{-}[Co(NH_3)_4(NO_2)_2]^+$
Croceo[d] trans(or 1, 6)-Dinitro- tetraamminecobalt(III)	Yellow or orange	$trans\text{-}[Co(NH_3)_4(NO_2)_2]^+$
Dichro trans(or 1, 6)-Dichloro- aquatriamminecobalt(III)	Green	$trans\text{-}[Co(NH_3)_3(H_2O)Cl_2]^+$

Color and IUPAC name	Color	Formula
Melano chloride	Black	A mixture of:

$$\left[\begin{array}{cc} (NH_3)_3 & (NH_3)_3 \\ & Co- NH_2 \rightarrow CoH_2O \\ Cl_2 & Cl \end{array}\right] Cl_2$$

$$\left[\begin{array}{cc} (NH_3)_3 & (NH_3)_3 \\ & Co - NH_2 \rightarrow Co \\ Cl_2 & Cl_2 \end{array}\right] Cl$$

and some

$$\left[\begin{array}{cc} (NH_3)_3 & \underset{Co}{\overset{NH_2}{\diagdown}} \underset{O_2}{\overset{}{\diagup}} Co & (NH_3)_3 \\ Cl & & Cl \end{array}\right] Cl_2$$

[a] Although the color terms were originally used to refer to specific cobalt complexes, they were later extended to designate types of compounds of cobalt or other metals. Thus the term 'luteo' has been used to refer to hexaammines, i.e. complexes in which the central metal atom is bonded to six nitrogen atoms, e.g. $[Co(en)_3]^{3+}$, where en = ethylenediamine, $[Cr(NH_3)_6]^{3+}$, $[Cr(en)_3]^{3+}$, $[Ni(NH_3)_6]^{2+}$, even though the colors of the compounds may be at variance with the color of $[Co(NH_3)_6]^{3+}$.

[b] These terms refer to the chloro compounds but may be used to refer to the corresponding compounds containing other halogens, e.g. bromopurpureo, $[Co(NH_3)_5 Br]^{2+}$, or bromopraseo, trans-$[Co(NH_3)_4 Br_2]^+$.

[c] These two ions are a classic example of *structural isomerism* (Kauffman, 1973c), which results from differences in the arrangement of atoms or groups of atoms in the complexes. In other words, an actual difference in bonding (*constitution*) exists between the different isomers.

[d] Strictly speaking, these terms refer to the ammonia (ammine) compounds, but they may also be used to refer to the corresponding ethylenediamine (en) compounds.

2

Alfred Werner, Nobel Laureate in Chemistry, 1913

Occasionally, one man may play such a central role in a particular field of science that his name becomes synonymous with that field. Alfred Werner, the undisputed founder and systematizer of coordination chemistry, is just such a man (Kauffman, 1976a). Even today, more than a half-century after his death, coordination compounds, particularly metal–ammines, are still known as Werner complexes, and the coordination theory is colloquially called Werner's theory.

At the time of its proposal in 1893 by a 26-year-old *Privat-Dozent**, this revolutionary theory rested upon a minimum of experimental data. Werner devoted his entire scientific career to the amassing of the experimental evidence required to prove the validity of his youthful assumptions. Beginning with a study of the hitherto unexplained 'molecular compounds' (metal–ammines, hydrates and double salts), his ideas soon encompassed almost the whole of systematic inorganic chemistry (Kauffman, 1967b, 1973a, 1973d, 1974f) and also found application in the organic realm. His experimental and theoretical papers remain even today a foundation and guide for investigations in coordination chemistry. He was the first to demonstrate that stereochemistry* is a general phenomenon not limited to carbon compounds, and it is no exaggeration to declare that his coordination theory has exerted an effect on inorganic chemistry comparable to that exerted on organic chemistry by the structural ideas of Kekulé, Couper, Le Bel and Van't Hoff.

In 1913 Werner received the Nobel Prize in chemistry, the first Swiss chemist to attain this honor. Although he was chosen specifically for his monumental work on coordination compounds, the implications and applications of his research extend far beyond the confines of inorganic chemistry. Even before he began his extensive series of researches on 'molecular compounds', he was vitally concerned with one of the most

† This account of Werner's life and work is adapted from Kauffman, 1966c.

basic problems of chemistry — the nature of affinity and valence. Coordination compounds provided him with a challenging and exciting means to this end. The true nature and extent of his achievement is perhaps best expressed in the words of the Swedish Royal Academy of Sciences which awarded him the Nobel Prize in recognition of '*his work on the linkage of atoms in molecules*, by which he has thrown fresh light on old problems and opened new fields of research, particularly in inorganic chemistry' [italics added].

Today, when the practical and theoretical significance of modern structural inorganic chemistry is unquestioned, it is clear that the foundations of this field were erected largely by one man — Alfred Werner — who, for this reason, is sometimes called 'the inorganic Kekulé'. We shall now examine Werner's life and work.

The genealogy of the Werner family can be traced back as far as the sixteenth century. The oldest progenitor known with certainty is Hans Werner dit Bötz, a pastor at Eschbach/Baden, Germany, who died in Mulhouse (Haut-Rhin) in the province of Alsace in 1612. During succeeding years, members of the family resided in various places, but they never strayed far from the fertile and picturesque plain of the Rhine, which forms a natural boundary between France and Germany.

Hans-Urban Werner (born 8 March 1784 in Siegen-Oberlauterbach (Bas-Rhin)), Alfred's paternal grandfather, was a farmer who later settled in Mulhouse, where he died on 16 May 1870. His marriage to Catherine Gerhardtstein (1782–1854), also of Siegen, resulted in six sons and one daughter. One of these sons, Jean-Adam Werner (born 18 September 1820 in Siegen; died 26 March 1893 in Mulhouse), a locksmith and later foundry worker and Alfred's father, married Barbara Léger of Oberseebach (Bas-Rhin), on 7 October 1850. The childless marriage ended on 14 June 1854 with Barbara's death, and on 6 August 1857, Jean-Adam Werner remarried. His second wife, Salomé Jeanette Tesché (born 9 January 1825 in Molsheim (Bas-Rhin), France; died 1 March 1903 in Mulhouse) was a daughter of Ferdinand Tesché, originally from Remscheid, Germany, and Jeanette Jetter, originally from Freudenstadt, Germany. Of this union were born a short-lived daughter, Adèle (26 September 1858–2 October 1858), and three sons, Adolf (13 September 1860–18 September 1908), Jules (16 February 1862–1 May 1869) and Alfred. Only Adolf and Alfred lived to maturity.

The founder of coordination chemistry was born in Mulhouse at midnight on 12 December 1866. During this same year, another Alfred — Alfred Nobel — invented dynamite and began to amass the huge fortune, which after his death was to be used for the prizes that mark the ultimate achievement in chemistry, physics, medicine, literature, peace and, later on, economics. Forty-seven years later, almost to the day, Alfred Werner was to receive the prize in chemistry.

Alsace, Werner's birthplace, has long cherished its independence, but of all the cities of this region, none valued its freedom more highly than Mulhouse. As a result of the Treaties of Westphalia in 1648 and the subsequent territorial usurpations of Louis XIV, by 1681 all of Alsace had become incorporated into France – with the sole exception of Mulhouse. Not until 1798 did this tiny republic, for economic reasons, voluntarily seek union with France. It was in this city of fiercely self-reliant and militantly autonomous citizens that Alfred Werner was born and raised. The year of his birth, 1866, was the year of the Seven Weeks' War against Austria, which decided the hegemony of Prussia in Germany. Four years later the Franco-Prussian War began, bringing with it events that made a deep and lasting impression on young Alfred during his formative years.

When Alsace was annexed to the second German Reich in 1871, more than 50 000 Alsatians, deeply French in spirit, chose to emigrate to France rather than to remain under the dominion of the Germans. The Werner family decided to remain in Mulhouse, but their sympathies remained entirely with France. Although by Bismarck's unpopular decree German was the official language, French remained the language spoken in the Werner home. Despite Werner's great reverence for German science, his political and cultural ties bound him to France. The spirit of rebellion and resistance to authority, so much a part of Werner's childhood, may well have contributed to the revolutionary and iconoclastic character of the theory with which his name is associated.

Like most of the Werners, Alfred's father was Catholic. His mother was originally Protestant but had been converted to Catholicism. Accordingly, at the age of six, young Alfred was enrolled at the Ecole Libre des Frères (Brüderschule), which he attended until his 13th year. During these early school years, the dominating traits of Werner's personality – a remarkable self-confidence and a stubborn independence which made it impossible for him to submit blindly to authority – already became evident. He was not overly fond of school and often played truant. Whenever a paddling was imminent, he delighted in outwitting the brothers by hiding a piece of cardboard in his trousers. Yet his remarkable intellect was so obvious that once when he had to sit on the last row in a class seated according to scholastic achievement, his teacher said to him, 'You could be first if you wanted to!'

Following his graduation from the Ecole des Frères in 1878, Werner entered the Ecole Professionelle (Höhere Gewerbeschule), a technical school. He attended this school, which is no longer in existence, until 1885. It was probably at about this time that Werner's interest in chemistry took firm hold, and he began experimenting at home.

If we glance ahead into Werner's later day-to-day notebooks, we find scientific data and personal notations intermingled with an utter disregard for the traditional dichotomy between professional and personal life. In his mature years, Werner was corpulent, and a daily record of his weight during a period of dieting is thus made with the same systematic thoroughness that characterized his scientific work. Even resolutions, which most of us usually make and break in a casual manner, were carefully committed to writing, for example, 'I won't buy myself any cigars until Christmas. – Alfred'.

Paul Karrer, Paul Pfeiffer and others have compared Werner to August Kekulé, and the comparison is an apt one. Werner, like Kekulé, was in his early years interested in architecture, and one is tempted to speculate whether architects and chemists require similar talents in dealing with the structural problems of their respective disciplines.

Werner's earliest known original scientific work was written in Mulhouse at the age of 18, less than a month before he left for military service in the German army. He later recalled:

Without my own experimental data, I simply compiled a study of urea compounds which, in my youthful enthusiasm, I believed would reshape all of organic chemistry. . . . I went to the director of the Chemieschule in Mulhouse, Prof. [Emilio] Noelting . . . and showed him the work. He took it and told me to return in eight days. Punctually and in a hopeful mood I presented myself, but in spite of all the praise he accorded my work, he did not conceal from me the fact that I would not yet achieve a revolution in organic chemistry with this work and that I would have to study much more. I was satisfied to some extent with this [evaluation] and immediately asked him how long he thought I would need in order to become a professor. With a smile he answered that I would have to be patient for 7 or 8 years.

Professor Noelting's prophecy was fulfilled eight years later when Werner was called to the Universität Zürich as successor to Viktor Merz.

The paper, entitled 'Contribution de l'acide urique, de séries de la théobromine, caféine, et leurs derivés', reveals the discrepancy between young Werner's passionate, romantic enthusiasm for chemistry and his still inadequate training in the subject. Although its style is banal and its chemical thinking often unsound, this work, in its broad scope and daring attempts at systematization, foreshadows the intellectual heights which Werner was to reach only a few years later.

This youthful effort of Werner's abounds with pompous exaggerations, and from Noelting's marginal comments, of which 'A little more modesty would be appropriate' is a typical example, we can almost visualize the

older man cautiously attempting to encourage the young, impatient rebel
and yet urging him to temper his enthusiasm with restraint and propriety.
The manuscript, which clearly shows Werner's awareness of and unlimited
faith in his growing powers, ends with a jubilant burst of elation that
borders on braggadocio:

> This immense uric group has thus been developed with an extra-
> ordinary simplicity, and soon we shall have this area of organic
> chemistry arranged as orderly as few others are now arranged.

Werner's extravagantly rococo signature and the date, 15 September
1885, end the paper with an almost theatrical flourish.

Two weeks after the completion of this urea paper, on 1 October 1885,
Werner began his compulsory military duty in the German army as a one-
year volunteer (*Einjährig-Freiwilliger*). Considering his antagonistic feel-
ings toward Germany, the experience must have been a traumatic and
ambivalent one. Yet it had its advantages. He was stationed in Karlsruhe,
that charming and beautiful city in Baden noted for its public squares,
ornate fountains and stately monuments, famed as the site of the first
international chemical congress in 1860.

The recruit register describes Werner at this time as blond, slender
(5 feet, 9 inches tall) and clean-shaven. The bushy mustache, so much a
part of the mature Werner, is not yet in evidence. On 1 July 1886, he
became a lance corporal and was considered suitable material for a non-
commissioned officer.

While at Karlsruhe, Werner availed himself of the opportunity to study
organic chemistry at the Grossherzogliche Technische Hochschule (now
the Technische Hochschule Fridericiana). As we shall see, it was only in
1892, shortly before his formulation of the coordination theory, that he
showed any deep interest in inorganic chemistry. Soon after leaving
Karlsruhe, Werner was to go to Zürich, Switzerland's largest city, where he
was destined to spend the rest of his life. Yet Mulhouse remained close to
Werner's heart, and the city reciprocated by deciding on 6 September
1965 to name a new street near the École de Chimie 'Rue Alfred Werner'
in his honor.

On 1 October 1886 Werner was officially discharged from the German
army, and less than three weeks later he established his first residence in
Zürich near the banks of the Limmat just down the hill from the Eidgenössisches Polytechnikum where he enrolled for the winter semester
1886/87. Within a remarkably short time, he felt completely at home in
his adopted city, and he quickly became proficient in the local patois,
Züri-Dütsch, perhaps understandably, for there are similarities between it
and the Alsatian dialect of Mulhouse.

The Eidgenössisches Polytechnikum (since 23 June 1911 known as the Eidgenössische Technische Hochschule — the Federal Institute of Technology or ETH) was then, as it is today, one of the foremost technical schools in the world. As the graduate of a foreign, non-accredited industrial school, Werner did not possess the Swiss *Maturitätsausweis* (maturity certificate), which permits a student to matriculate at any Swiss university without taking an entrance examination. The results of his examination are so revealing as to merit our detailed consideration. The grade in chemistry was, not unexpectedly, 6 (this being the highest and 1 the poorest grade), and the grades in most of the other subjects were satisfactory (drawing, 4, 5; natural sciences, 4 1/2; physics, 4 1/2; composition, 5; political and literary history, 6; and French, 6). The grades in mathematics, however, were 2 and 4 1/2. Even more surprising, in view of Werner's amazing ability to conceive things structurally, are his failing grades (1 1/2 and 2) in descriptive geometry. His deficiency in mathematics is confirmed by the grade of 2 which he received in 'Higher Mathematics', a course taken during his first semester (winter, 1886/87). In view of his obvious weakness in this area, it is not surprising that throughout his entire career Werner's contributions were essentially of a qualitative nature — even his celebrated conductivity* studies with Miolati (1893, 1894) were actually only semi-quantitative.

Today, coordination chemistry, like most branches of modern science, is rapidly becoming more mathematical and abstract, a trend which will probably accelerate in the future. The power and advantages of such a mathematical approach are unquestionable. Yet we should never forget that the founder of coordination chemistry, a typical example of a non-quantitative genius, once failed mathematics in school. For those who are excessively preoccupied and enamored with a quantitative approach, Werner should provide a dramatic proof that mathematical ability is not the only prerequisite for success in chemistry.

After passing his oral and practical examinations (October 1888; July 1889), Werner reached his first academic goal. On 3 August 1889 he was awarded the degree of *Technischer Chemiker* and became an unsalaried assistant (*Hilfsassistent*) in Georg Lunge's (1839–1923) chemical-technical laboratory (1889–1890). While still *Hilfsassistent* to Lunge, Werner began work on his doctoral dissertation under Arthur Hantzsch's (1857–1935) direction. Despite Hantzsch's more than 500 publications, his greatest discovery was probably Alfred Werner, who was not only his most outstanding pupil but also his lifelong friend. Werner always regarded Hantzsch as the outstanding influence in his early career, and he dedicated his first book, *Lehrbuch der Stereochemie* (1904), to his former teacher.

It was in the span of three short but eventful years (1890–1893) that Werner produced his three most important theoretical papers — 'three giant steps', as his former student and assistant Robert Huber so aptly called them. The first of this illustrious trio was his four-part inaugural dissertation, 'Über räumliche Anordnung der Atome in stickstoffhaltigen Molekülen' (1890). The first or theoretical part, which appeared under the joint authorship of Hantzsch and Werner (1890) even before the dissertation itself was printed, was not only Werner's first publication but still remains his most popular and important work in the organic field (Kauffman, 1966a, 1972a, 1976f). In this paper Werner and Hantzsch, by transferring the Le Bel and Van't Hoff concept of the tetrahedral carbon atom to the nitrogen atom, were able to explain a great number of puzzling cases of geometric isomerism. For the first time, the stereochemistry of nitrogen compounds was placed on a firm and satisfactory theoretical basis. Although he continued to publish occasionally on organic topics throughout his scientific career, Werner's attention soon shifted to inorganic chemistry.

In spite of early attacks on the theory by Victor Meyer and Karl von Auwers and later attacks by Eugen Bamberger and other chemists extending several decades into the present century, Werner and Hantzsch's view has withstood the test of time, and today, with only slight modification, it takes its rightful place alongside the Le Bel and Van't Hoff concept of the tetrahedral carbon atom as one of the cornerstones of stereochemistry. (See Ramsay, *Stereochemistry*, in this series.)

In July of 1890 Werner's dissertation was 'approved very favorably', and on 13 October 1890 the 23-year-old Alfred Werner was awarded the degree *Doktor der Philosophie* 'with special recognition of superior performance'.

Werner now decided upon an academic career, the first step of which was to become a *Privat-Dozent**, a position which required the *venia legendi* or *venia docendi*, the privilege of lecturing at a university, which was awarded only upon acceptance by the faculty of a *Habilitationsschrift**, a paper embodying the results of original and independent research. With his characteristic single-minded determination, Werner set about compiling such a paper. During the busy years 1890 and 1891, he had been conducting his own research on benzhydroxamic acid derivatives in the analytical laboratories of the Polytechnikum, and these results served as the experimental and subordinate part of his *Habilitationsschrift* (1891).

In the first and theoretical part of his *Habilitationsschrift*, 'Beiträge zur Theorie der Affinität und Valenz' (1891), the fledgling doctor of philosophy chose to attack no less than the supreme patriarch of structural organic chemistry, August Kekulé himself. In this, his second 'giant step' in stereochemistry, Werner attempted to replace Kekulé's concept of rigid, directed valencies with his own more flexible approach, in which he viewed affinity as a variously divisible, attractive force emanating from the center of an atom and acting equally in all directions (Kauffman, 1979 c, e). By the use of this new concept and without assuming directed valencies, Werner was able to derive the Van't Hoff configurational formulas. Although this paper contains the seeds which later were to flower forth in the primary valence* (*Hauptvalenz*) and secondary valence* (*Nebenvalenz*) of the coordination theory, it deals exclusively with organic compounds.

Werner was aware of the serious nature of his attack upon the current, firmly entrenched valence theory, and he fully expected strong resistance to his new ideas. But the anticipated attack did not take place, at least at that time. Even before the entire *Habilitationsschrift* was printed, this important paper was published separately in a journal of limited circulation, the *Vierteljahrsschrift der Zürcher Naturforschenden Gesellschaft* (1891), where it elicited little notice until brought to the attention of the scientific world by a discussion of its concepts in Werner's *Lehrbuch der Stereochemie* (1904).

On 16 October 1891, from the Hotel Pfauen where he now resided, Werner submitted his recently completed *Habilitationsschrift* to the Hohe Schweizerische Schulrat and petitioned them for the *venia docendi* in chemistry at the Polytechnikum. He did not sit back and wait or worry about its acceptance but almost immediately departed for the Collège de France in Paris where during the winter semester 1892/93 he pursued research in thermochemistry under the direction of the illustrious French physical chemist, historian of chemistry and statesman Marcellin Berthelot (1827–1907).

On 4 January 1892, while Werner was still at work in Paris with Berthelot, the Hohe Schweizerische Schulrat, which had accepted his *Habilitationsschrift*, named him *Privat-Dozent* in the *Freifächerabtheilung* (elective subjects division) of the Polytechnikum. For the subject of the public inaugural address (*Antrittsvorlesung*) required of every new *Privat-Dozent* sometime during his first year of teaching, Werner chose the perenially popular 'benzene problem', which he had already touched upon in his *Habilitationsschrift*. In this lecture, 'Kritische Beleuchtung der heutigen Benzoltheorie', which was delivered during the summer semester 1892, he first reviewed and called attention to the inadequacies of the various structural formulae proposed for benzene by Kekulé, Claus, Loschmidt, Thomsen, Sachse and other chemists. He then presented his own views, which had already been developed in the *Habilitationsschrift*. By assuming that affinity is a force transmitted by a process akin to the emission of light, he showed how some of the atoms in the benzene ring would be brightly illuminated while others would be placed in the shade. This clever analogy permitted him to explain many experimentally observed aspects of aromatic substitution – again without invoking directed valence forces (Kauffman, 1967d, e, 1979c, e).

Werner's first course, 'Atomic Theory', was given at the Polytechnikum during the summer semester 1892 (19 April– 6 August). In the early years of his teaching career, he was handicapped by factors common to novice instructors – youth and inexperience. In addition to intellectual gifts, it took extreme tact, a character trait for which Werner was never noted, to build and preserve good relations with both students and colleagues. At this time, he grew a mustache and beard so as to appear older and more authoritative to his students, who were only a few years younger than he. The beard soon disappeared, but the inseparable mustache remained with him throughout his lifetime.

Werner's career at the Polytechnikum was not a long one, for on 29 September 1893 he left to accept a call to the Universität Zürich. The call came about largely because of the almost overnight fame which Werner had received as a result of the publication of his most important theoretical paper, his third and greatest 'giant step', 'Beitrag zur Konstitution anorganischer Verbindungen' (1893), in which he had proposed the basic postulates of his epoch-making coordination theory. Unlike his *Habilitationsschrift*, this paper did not appear in an obscure journal but in the third volume of the recently founded *Zeitschrift für anorganische Chemie*, where it aroused instantaneous interest – and criticism.

On 31 August 1893, Werner was officially appointed as *Ausserordentlicher (Extraordinarius) Professor für organische Chemie* at the Universität Zürich (then called the Zürcher Hochschule) for the usual term of six years, with a salary, in addition to *Kollegiengelder* (fees paid by students), of Sw. Fr. 2500. He was to be entrusted with 'organic chemistry, along with supplementary special lectures, theoretical chemistry, and Laboratorium A'. Thus, with the advent of the winter semester 1893/94 (17 October 1893– 10 March 1894) Werner began his long and distinguished career at the Universität Zürich.

In his coordination theory, Werner discarded the confining rigidities of the Kekulé valence theory with its artificial distinction between 'valence compounds' and 'molecular compounds' in favor of a new and revolutionary, comprehensive approach in which the constitution of metal–ammines, double salts and hydrates were viewed as logical consequences of a new concept – the coordination number (*Koordinationszahl*). By use of this unifying concept, he divided metal–ammines into two classes – those with coordination number six and those with coordination number four. For compounds of the first class, he postulated an octahedral configuration and, for those of the second class, a square planar or tetrahedral configuration. He then proceeded to demonstrate the correctness of these stereochemical views by citing various reactions, transformations and cases of isomerism. In this classic paper, Werner did not limit himself exclusively to the constitution and configuration of 'molecular compounds', but he also speculated upon other topics such as the state of metal salts in solution and the polarization effects involved in chemical bonding.

The coordination theory not only provided a logical explanation for known 'molecular compounds' but also predicted series of unknown compounds, whose eventual discovery lent further weight to Werner's controversial ideas. He showed how ammonia could be replaced by water or other groups, and he demonstrated the existence of transition series between ammines, double salts and hydrates. Werner recognized and named many types of inorganic isomerism such as coordination isomerism, polymerization isomerism, ionization isomerism, hydrate isomerism, salt isomerism, coordination position isomerism and valence isomerism (Kauffman, 1973c). He also postulated explanations for polynuclear complexes (Kauffman, 1973b), hydrated metal ions, hydrolysis, and acids and bases (Kauffman, 1973d).

'The inspiration [for the coordination theory] came to him like a flash', related Paul Pfeiffer (1928), drawing on his many conversations with Werner. 'One morning at two o'clock he awoke with a start: the long-sought solution of this problem had lodged in his brain. He arose from his bed and by five o'clock in the afternoon the essential points of the coordination theory were achieved'. We shall have more to say about the genesis of the coordination theory in Chapter 6. For the present we shall try to account for the 'the flash of genius' that gave birth to this highly original theory.

The force of Werner's complex and often inconsistent personality, of which we can catch only striking glimpses here and there from the limited material available to us, undoubtedly harbors the key to this riddle of the act of creation. Highly ambitious, dauntless in the face of rebuffs and failures, egocentric, imaginative, intuitive, aggressive, passionate, sensitive, impulsive, obsessed with self-imposed tasks – all these traits known to be associated with high creativity could be used in describing Werner. Add to this a high degree of native intelligence and a childhood spent in an atmosphere of political unrest, rebellion, and resistance to and criticism of authority, and we have the highly combustible mixture which ignited in that sudden nocturnal explosion, the creation of the coordination theory. Werner himself attributed his theory to 'a strong feeling of independence . . . a lack of belief in authority . . . and an urge toward the truth'.

According to Pasteur, 'chance favors the prepared mind', so we might do well to try to determine how long Werner had mulled over in his mind the puzzle of 'molecular compounds' before the brilliant systematization of these compounds came to him in one blinding flash of visual insight (Kauffman, 1976e). Again we turn to Pfeiffer: 'When [Werner], in the course of working out a theoretical-chemical lecture, became absorbed in the prevailing theories of metal–ammonia salts and related compounds, he soon became convinced that conventional valence theory could not completely explain the constitution of these compounds' (Pfeiffer, 1920b). We know that Werner did not begin his teaching career until the summer semester of 1892 (19 April–6 August), so the lecture cited by Pfeiffer could not have taken place more than a mere six or seven months before the coordination theory was submitted to the *Zeitschrift für anorganische Chemie* (December 1892). But could Werner have become interested in metal–ammines before the lecture in question?

Possibly, but not likely. Werner was trained as an organic chemist. All his previous publications, with the exception of one, dealt with strictly organic topics, and there is no indication that he ever evinced anything more than passing interest in the inorganic field. Although it cannot be proved, it is thus extremely likely that Werner's first interest in inorganic chemistry that was more than superficial arose in connection with his course 'Atomlehre' taught during the summer semester of 1892. If, on the contrary, it did not arise until Werner first taught 'Ausgewählte Kapitel der anorganischen Chemie' (winter semester 1892/93 (10 October 1892–18 March 1893)), his achievement is all the more amazing – almost miraculous!

Yet Werner's case is far from unique, and many more examples can be found, not only in chemistry but in other sciences as well. The iconoclast, almost by definition, is young and inexperienced. To quote the contemporary historian of science, Thomas Kuhn (1970):

> Almost always the men who achieve these fundamental inventions of a new paradigm have either been very young or very new to the field whose paradigm they change . . . . Obviously these are the men who, being little committed by prior practice to the traditional rules of normal science, are particularly likely to see that these rules no longer define a playable game and to conceive another set that can replace them.

Once again, we shall succumb, as have others, to the irresistible temptation to compare and contrast Werner with Kekulé. Both men, with the intuition of genius, brought order into large fields of chemistry, Werner into inorganic and Kekulé into organic. Whereas Kekulé, however, was primarily a strong theorist and made no major contributions as an experimentalist, Werner was not only the founder of coordination chemistry but also the greatest experimenter in this field. Almost every aspect of modern coordination chemistry, if traced back far enough in time, leads to some experimental work of Alfred Werner.

The state of inorganic chemistry during the latter half of the nineteenth century was undistinguished, compared to that of organic chemistry. Only a few individual figures stood out clearly, notably Moissan, Ramsay and Werner – in sharp contrast to the many top rank organic chemists. A long stalemate had resulted from excessive dependence on organic structural concepts. Not until the end of the century, as a consequence of Werner's work, did structural inorganic chemistry take a profitable direction.

The central role played by Werner during these crucial years and the glowing success of his labors in coordination chemistry sometimes tend to obscure the fact that Werner, especially during his early years, made many worthwhile contributions to organic chemistry. As we have seen, he was originally called to Universität Zürich to teach organic chemistry, and it was not until the winter semester 1902/3 that he was finally assigned the main lecture course in inorganic chemistry, which he continued to teach along with organic chemistry throughout his career. Forty-five of his 174 publications dealt with organic themes such as oximes; hydroxamic and hydroximic acids; phenanthrenes; hydroxylamines; azo, azoxy, hydrazo and nitro compounds; and dyestuffs (Kauffman, 1976f). His investigations and interpretations of the Walden inversion* are still of value today.

For half a dozen years following his call to the university, then, Werner's attention was divided between organic and inorganic chemistry. Of his first 30 publications (1890–1896), the organic papers outnumber the inorganic ones by a ratio of two to one. Not until 1898 when his reputation in coordination chemistry had been firmly established did the number of his inorganic papers (21) reach that of his organic ones.

It was a time when organic chemistry was in a state of extraordinary development, overshadowing the other fields with its brilliant triumphs. Werner, attracted by the promise of early fame in organic work, yet now basically drawn to inorganic chemistry, wavered for a time, troubled by ambivalence and plagued with doubts.

'Several times I was at the point of again turning completely to organic chemistry, a field in which I would certainly receive more recognition with less work', Werner confided to Hantzsch in a letter of 25 November 1897. 'Again and again I have felt compelled to return to inorganic chemistry, perhaps because I really overestimate the importance of these investigations'. By 12 July 1899 Werner had made his decision. 'This periodic vacillation is now overcome', he announced to Hantzsch. 'Inorganic chemistry presents me with so many problems whose solutions attract me that I shall definitely take the path in the inorganic direction; also, I hope to be able to achieve more there than in the organic field.'

Despite his almost total commitment to inorganic chemistry, Werner continued to enrich the organic literature with 21 more papers. Yet he now considered himself an inorganic chemist, and, with the zeal typical of a proselyte, he did everything within his power to advance the cause and status of his newly chosen field.

Only two years after his appointment, the *Erziehungsrat* concluded that the rank of *Extraordinarius Professor* was not commensurate with Werner's importance, inasmuch as his outstanding qualities, increasingly recognized by prominent European scientists such as Emil Fischer and Adolf von Baeyer, had already resulted in a call to an important German university. This call Werner refused, and in appreciation of his decision as well as in recognition of his outstanding teaching and research activity, on 8 June 1895 he was promoted to *Ordinarius Professor*.

Another call, this time from Berne, was refused in 1897 and led to another increase in salary and to the first of many promises for improvements in the laboratory facilities. This pattern of a call followed by a salary increase or improvements in the institute was to be repeated many times during Werner's career as the number of tempting offers multiplied — Vienna (1900), Basel (1902), Eidgenössisches Polytechnikum (1905) and Würzburg (1910). At the same time, he was awarded numerous honorary memberships and degrees by European and American universities and scientific societies.

Not long after 'Beitrag zur Konstitution anorganischer Verbindungen' had assured him a reputation as a scientist of some consequence and after his lectures at the Universität had proved him to be a professor fulfilling the promising predictions made about him by his teachers, Werner committed himself to Zürich, in which he now felt thoroughly at home, in two decisive ways: he married a Swiss woman, and less than a month later, on 24 October 1894, he was granted the rights of local (*Bürgerrecht*) and federal (*Landrecht*) Swiss citizenship. Although it was possible to maintain dual citizenship, Werner chose to renounce his German citizenship, a not unexpected course of action in view of his antagonism toward Germany.

Werner's bride of 1 October 1894, Emma Wilhelmine Giesker, was the adopted daughter of Ernst August Giesker, a Protestant pastor. She was 21 (born 14 December 1872 in Zürich-Enge) when she met the young and dashing *Extraordinarius Professor*, while both were out horseback-riding one day. After a brief courtship and engagement of only a few months, the couple were married. A son, Alfred Albert Julius, nicknamed Fredy, was born on 22 July 1897, and on 16 April 1902 the Werner's second and last child, Johanna Emma Charlotte, was born.

Unfortunately, but perhaps inevitably, Werner's youthful involvement and pleasure in his home and family was to be displaced by an increasing and intense absorption in his scientific work. As the years passed, the image that we get of Werner, as described by colleagues, former students and friends, is that of a man who lived almost exclusively for his science and who slept little; in the evening he was rarely at home but was either at the institute or with his friends in intensely compressed hours of relaxation which involved, at one time or other, chess, billiards, bowling and the Swiss national card game of *Jass*. Most pathetic of all was his increasing and well-known dependence on alcohol, which Werner, with his characteristic truthfulness, was the first to admit.

Inevitably, too, Emma Werner's increasing preoccupation with her home and children, an absorption fostered by the *Kinder–Küche–Kirche* tradition of Protestant Zürich, contributed to Werner's alienation from his home. As Werner's horizon expanded and as his professional stature increased, his wife was becoming more and more absorbed in the minutiae of housekeeping. She abandoned her early interest in painting and other extrafamilial activities. In addition, contemporaries report that the pair were quite different in terms of personality; socially, Frau Werner was quiet, somewhat withdrawn, and not nearly as adept as her extroverted, jovial husband, who, in this area too, was as intense, skillful and energetic as he was about everything that he enjoyed.

When Werner came to the Universität, its Chemical Institute was housed in the old building at Rämistrasse 85, which had served as the Chemical Institute of the Polytechnikum until the latter had been moved into new facilities at Universitätstrasse 6 in 1887. Although considered ideal at the beginning of the 1860s, the building, which also housed the chemical laboratories of the *Kantonsschule*, the *Kantonschemiker* and the Hygiene and Pharmacology Institutes of the university, proved inadequate even during Victor Meyer's tenure from 1872 to 1885. The laboratories were much too small to accommodate the burgeoning enrollment and were completely incapable of meeting the technical and sanitary demands placed upon them.

Werner's students worked in what they aptly nicknamed the 'Catacombs' (*Katakomben*) – unfinished cellars and storage rooms for wood, so poorly illuminated that artificial gas lighting was required even at noon (Plate 2). The unhealthy contrast of steam pipes overhead and cold cement floors below, along with the penetrating reek of pyridine (there was no ventilation to speak of) completed the dismal scene.

Yet Hermann Helmholtz's dictum that 'the best works come out of the worst laboratories' may have some degree of validity, for it was in the Catacombs that the major portion of Werner's life work was performed. In no way did the substandard conditions seem to dampen the youthful

Inorganic Coordination Compounds

Plate 2. The 'Catacombs', Old Chemical Laboratory, Universität Zürich
[Kauffman, 1966c, p. 65]

high spirits, dedication and sense of humor of those to whom Werner later referred as his 'enthusiastic young fighters in the battle for knowledge'.

Many of these students were foreigners, attracted by Werner's growing international reputation. The Swiss students had at their disposal the excellent facilities of the Polytechnikum, and quite understandably, few chose to work at the university laboratories. Much of the early research, Werner admits, could not have been accomplished without the dedicated participation of students, many of them women, who flocked to the institute from all parts of the world – Germany, Russia, Poland, Austria, Hungary, Bohemia, England, Holland, Romania, Bulgaria and the United States, to mention some of the countries represented. The laboratories were so swamped with Russian students, many of them fleeing the harsh conditions of the Czarist regime, that one *Weihnachtskommers* Zeitung* announced that 'beginning next semester Prof. Nunwiegehts* will lecture in Russian. The few students who are not conversant with this language should arrange to take private lessons'. The late eminent chemical historian John Read, who received his doctorate under Werner in 1907, described the cosmopolitan atmosphere of Werner's laboratory as an education in itself.

Year after year, with his characteristic persistence, Werner continually pressured the Zürich authorities to provide improvements in the laboratory conditions. Finally, on 20 February 1905, the *Kantonsrat* granted a

request of the *Kantonsrätliche Kommission* for Sw. Fr. 1 400 000 to be
used in the erection of a new institute, and this request, together with
other increases in Zürich's contributions to education, was passed by a
vote of 31 436 to 15 193 in a referendum held on 25 June of that year.

Werner's joy at the decision, understandably, was immense. John Read
recalled how the new institute was the major theme of the 1906 *Weihnachtskommers* held at the Casino Hottingen and how Werner, in top
oratorical form, described the wonders of the modern structure as part of
his traditional annual speech. 'It will have water, air, steam and vacuum
connections, but one more thing will be provided without fail— and that
will be connections for — beer!'

Initially, the completion of the new building was scheduled for the
fall of 1907, but various circumstances prevented the realization of this
objective until 1909. For Werner, the construction period meant innumer-
able time-consuming consultations and negotiations with officials, architects, plumbers and carpenters as well as seemingly endless and often dis-
appointing compromises and aggravating delays.

The building was finally completed, and on 27 February 1909 Werner's
students began to move their work benches and laboratory equipment
into the new building. In one of those petty misunderstandings that sometimes disturb the calm waters of academic life, *Kantonsbaumeister*
(cantonal architect) Fietz complained of their misbehavior, and Werner
quickly sprang to their defense in a four-page reply (10 March 1909) to
the accusation. The letter, in which Werner agrees to compensate for the
minor damages, gives us a glimpse of the jubilant mood at that time of
both Werner and his *Doktoranden**:

> As far as the so-called parade is concerned . . . the students
> hired four musicians who played a sad farewell to the old build-
> ing and then, marching in front of the automobile (which
> brought the things to the new institute), played a gay march.
> All who witnessed this parade considered it a highly successful
> student joke. Thus, in this regard too, the severe word 'mis-
> behavior' can hardly be used.

Today the massive stone structure at Rämistrasse 76 still houses the
Chemisches Institut der Universität Zürich. It stands as a monument to
the energy, enthusiasm and foresight of Alfred Werner.

All reports from those who actually heard Werner lecture during his
prime are unanimous in their glowing descriptions. They differ only in
their choice of superlatives. Enthralling! Inspiring! Magical! Fascinating!
A perfect joy! His voice has been most often described as sonorous. His
delivery was unusually clear, calm and precise. Its pace was comparatively

slow so that his German was easy for even foreign students to follow and for all to transcribe into notes, a quality which surely filled his listeners with gratitude. His inner warmth, fire and contagious enthusiasm caught his audience and carried them along with him.

As Werner's fame as a lecturer spread beyond the confines of the *Philosophische Fakultät II* (Mathematics–Natural Science Faculty), students from other disciplines, even those as far afield as theology and law, would attend his lectures, drawn by his magnetic personality. Many of the medical students decided to become chemists. Even the spacious auditorium of the new institute was soon filled to overflowing, and the crowded conditions of the old institute were repeated. In the winter semester 1913/14, during which Werner received the Nobel Prize, 336 students eager to hear Werner's course on inorganic chemistry, squeezed into an auditorium with a seating capacity of 209. An excerpt from a *Weihnachtskommers* catechism drolly comments upon the situation:

Question: When does a chemist occupy the minimum volume?

Answer: In Professor Nunwiegehts'* lecture.

It was characteristic of Werner never to demand of others what he did not demand of himself. His first lecture of the day began at 8:15 a.m., and he was always on time, even after the late hours spent with friends at his favorite drinking spots, the Seehof or Pfauen. He never showed any effects of the previous evening's conviviality. The only recorded instance of his being late to a lecture was on the occasion of the first resolution of a coordination compound (see p. 126). In this case the lecture was canceled at the last moment.

A search for information on Werner's relations with his students immediately uncovers inconsistencies and conflicting testimony. In interpreting the available data, we should take all the stories with a grain of salt and remember that students are often prone to rationalize their own inadequacies by criticizing their teachers.

Especially contradictory are the stories concerning Werner's behavior during examinations, all of which were, of course, oral. When it came to intellectual weaknesses, Werner's humor could become grim and even sadistic. Various tales were commonly circulated and possibly exaggerated among students and faculty, and Werner's reputation as a difficult and intimidating examiner, whether justified or not, became rather firmly established. Yet the superior, conscientious students found Werner to be consistently kind, calm, supportive and quite fair in the questions which he asked.

In view of the extreme divergence in the picture of Werner's relations with students, we would do well to consider the alleged incidents in the light of two factors — the type of student involved and the time in Werner's life when the incident supposedly occurred. Even in his early years Werner possessed the moody and temperamental nature that we have, rightly or wrongly, come to associate with the artist rather than the scientist. Particularly in his later years, as his responsibilities increased and his illness caused him immeasurable and untold physical and spiritual anguish, naturally the more impulsive and impatient side of his character came to the fore.

Perhaps Werner himself best summed up his attitude in describing himself as one who 'always preferred to place the factual above the personal'. His demands on students were based entirely on the demands of the subject matter. Any consideration of the ability or interest of the student was to him completely beside the point.

Werner kept close control over the undergraduate laboratories and habitually made the rounds twice a day, once in the morning and once in the afternoon. However, he rarely stopped at the bench of a student; a brief discussion with the young assistant in charge was the limit of Werner's activity in these laboratories. Professor Paul Pfeiffer proved to be a more approachable source of information and advice to undergraduates. Most of Werner's time went to the fantastic number of doctoral candidates doing research under his direction.

The attainment of a doctor's degree at the Universität Zürich required a difficult course of research and study, but once a student was accepted by Werner as a *Doktorand* or *Doktorandin*, he or she was fairly certain to complete the work successfully. We have already seen how virtually Werner's entire research career was based on one great intuitive prediction, the validity of which was so firmly established in his mind that he did not hesitate to invest a lifetime of work in proving it. In the same way, a *Doktorand* was never assigned a project until Werner was certain of its theoretical basis. In addition, its practical feasibility was usually checked in a preliminary way and on a small scale by one of Werner's assistants before the problem was actually assigned to a student. Upon the completion of the research, Werner with his characteristic thoroughness did not merely accept a student's analytical results but had each one routinely repeated by his assistant (quite a job for 25–30 *Doktoranden*!).

Perhaps this is an appropriate place to mention the personal technique which Werner developed in his own individual experimental research. In these days when the trend toward instrumentation is increasing rapidly and when useful but elaborate instruments are in danger of degenerating into status symbols, it should be emphasized that Werner laid the experimental foundations of coordination chemistry by using the simplest

Plate 3. The Werner collection of coordination compounds (partial view)
[photograph by Herr Richard Taubenest, Universität Zürich]

kinds of physical and chemical equipment. Much as a composer may deliberately turn his back on the immense forces of the symphony orchestra in order to work in a more modest medium such as that of a sonata or a string quartet, so did Werner choose the most elementary, unpretentious and direct means that would give him the answers that he eagerly desired as quickly as possible.

In his private laboratory, Werner kept a small experiment table reserved for his personal use. On the table stood several microburners, microfilter supports and small platinum spatulas, but the majority of the table's surface was covered with what Pfeiffer has described as Werner's real tools — small hemispherical glass dishes filled with complexes of all the colors of the rainbow. Although none of these *Schälchen* were labeled, Werner could immediately identify the contents of each dish even after long periods of absence from the laboratory.

The fruit of a quarter-century's whirlwind of research activity on the part of Werner and his students is still preserved in a narrow storage room adjoining the *Grosser Hörsaal* at the Chemisches Institut der Universität Zürich. Here the browsing visitor can admire the literally thousands of preparations contained in carefully labeled tubes which are stored in more than 100 drawers that are housed in a gigantic, heavy, wooden cabinet which reaches almost to the ceiling (Plate 3). Werner explained his accomplishment quite simply and unpretentiously:

I am conscious of having worked quite diligently. But chemical work was always a pleasure for me, and I have experienced the purest pleasures in the laboratory, when on the basis of reflec- tions I arrived at new conclusions which could be confirmed experimentally.

It was in his *Schälchen* that Werner, armed with only a platinum spatula and the most primitive of apparatus, subjected the complexes, produced on a larger scale by his assistants, to the most diverse reactions, transformations and operations. He would qualitatively examine sub- stances by treating small amounts of them on watch glasses with different reagents, using porous clay plates to separate precipitates, and washing these precipitates by moving them with a spatula to a fresh portion of the plate where they were treated with various solvents. To observe Werner, with his finely jointed, slim, artistic hands, convert one compound into another or into an entire series of derivatives by this simple method was to realize that one was in the presence of a true master experimentalist.

As we have already seen, before assigning a topic to a student, Werner would have his *Privat-Assistent* carry out preliminary experiments in miniature, and only after he had convinced himself of the practicality of the proposed research did he allow full-scale work to proceed. While one may question the pedagogical value of this procedure, one cannot but admit its efficiency and predictability. Once a student was accepted by Werner as a *Doktorand*, he knew that his chances of attaining the degree were excellent, although he may have had little choice in his research topic and little opportunity to learn by making his own mistakes.

In this connection, one might also wish to ponder whether it was this high degree of regulation and supervision which may have prevented the formation of a Werner school in the usual sense of the word. It is true that among his students and one-time associates we encounter the names of academic and industrial researchers, but only a few of these men such as Paul Pfeiffer, Alexander Gutbier or Yuji Shibata earned their reputations in the field of coordination chemistry. Perhaps the impact of Werner's powerful, authoritarian personality and the impression of his control and mastery of his field deterred most of those who had worked with him from any thought of following in his footsteps.

Another reason may lie in the thoroughness, breadth and completeness of Werner's lifetime of research in coordination chemistry. Indeed, there is scarcely a single aspect of the field in which he had not performed some experimental or theoretical work. Ironically enough, Werner's contributions to the field were so enormous and comprehensive that for a number of years many chemists gained the false impression that nothing further remained to be discovered in the area. It is only in the last three decades that this misconception has been overcome and that we have witnessed a resurgence of interest in coordination chemistry. It is unfortunate that Werner did not live to see this renaissance.

On 12 November 1913, Werner received a terse telegram: 'Nobel Prize for chemistry awarded you. Letter follows – Aurivillius'. Through this brief wire from the Secretary of the Royal Swedish Academy of Sciences, Werner learned that he had become the 14th chemist to receive the internationally famous award and the first Swiss chemist to attain this honor. He reacted to this moment of ultimate triumph with typical candor: 'I had not completely eliminated the thought that it would come some day, but I hadn't expected it this year'.

The imposing award ceremony took place in the Grand Hall of the Royal Academy of Music in Stockholm at five o'clock on the afternoon of 10 December 1913, the 17th anniversary of the death of Alfred Nobel, with members of the nobility and high officials in attendance. After a brief speech by T. Nordström, President of the Royal Swedish Academy of Sciences, in praise of Werner's achievements, His Royal Highness King Gustav V officially presented the chemistry prize to Werner, who at 47 was the youngest of the recipients. Werner received a gold medal, an elaborate scroll, and the sum of 143 010.89 Swedish crowns (c. Sw. Fr. 197 000). According to the official certificate, Werner was awarded the prize

in recognition of his work on the linkage of atoms in molecules by which he has thrown fresh light on old problems and opened up new fields of research, especially in inorganic chemistry.

Werner delivered his Nobel Prize address, 'Über die Konstitution und Konfiguration von Verbindungen höherer Ordnung' the following day, 11 December (Werner, 1913). The 15-page lecture was and still is considered a model of compressed exposition, summarizing as it does the labor of 20 years and in addition discussing future problems. Again, as usual, Werner was careful to give credit where credit was due, as has been pointed out by one of his former students, Charles H. Herty:

It is interesting that this address was made in that part of Europe from whence came at one time such strenuous denunciations of his views, and it is pleasant to note that in his reference to one of the chemists of the northern country, Jørgensen, no scars have been left from the bitter controversy which prevailed in the earlier days.

In general, the tragic tale of Werner's final illness is well known. What is not widely known, however, is that this illness was preceded and foreshadowed by definite indications and warnings that might have been heeded by someone less compulsively dedicated to his science. It may come as a real surprise to many who knew him that the man who presented such a vigorous, active and powerful figure to the outside world suffered from nervous headaches and signs of chronic overwork throughout much of his life. But Werner was not one to complain or indulge in self-pity. Perhaps if he had moderated his intense activity, perhaps if he had refrained from 'burning the candle at both ends' year after year, his life might not have been cut short at the very zenith of his career.

Yet such speculation is pointless. Alfred Werner was a true *Arbeitsmensch*. For him, life without a complete and unrestrained pursuit of knowledge would have been no life at all. In a very real sense then, Werner died a victim of his work.

As early as 25 May 1899, two decades before his death, Werner confided to his close friend Arturo Miolati:

At the end of the winter semester, I already felt extraordinarily tired, which was not surprising, considering the constant overcrowding of the laboratory with *Praktikanten* and my consequently increasing work-load. Nevertheless, I had the unfortunate idea not to use the Easter vacation for a rest but instead to catch up with overdue work during this entire time and to prepare no less than twelve papers for publication.

No wonder the beginning of that summer semester was very difficult for Werner and that a severe case of pneumonia complicated by nervous strain led his physician to insist upon a lengthy period of complete rest.

Werner paid for the folly of his overwork of 1899. On 20 September 1900 he wrote to Hantzsch that he had not worked much that year, for he had suffered a severe attack of whooping cough, his nerves were troubling him, and he had been forced to spend five weeks at a hydropathic sanatorium. Through 1902, 1903 and 1904, he had to decline offers from publishers and journal editors because of ill health. For example, on 24 December 1903 he resigned from participation in the *Chemische Zeitschrift*, citing as reasons his lack of time and his severe nervous headaches. Although his physician attributed these headaches to overwork, Werner, unwilling to reduce his activities, did not agree with the diagnosis. In the ensuing years, the literature of chemistry was immeasureably enriched, but at a heavy price, which was paid by a doggedly determined man working in the laboratory on the Rämistrasse.

Shortly after he had received the Nobel Prize, the final difficult trial for Werner and his family began. The dark shadow of the progressive, degenerative and fatal illness diagnosed as 'general arteriosclerosis, especially of the brain' started its inexorable inroads upon a body already weakened by years of overwork and excessive consumption of alcohol. His physician's order to abstain from his beloved, habitual cigar was only the first and the smallest of the many sacrifices and changes in his way of life that Werner would be called upon to make.

On 8 July 1915 Werner asked the *Zürich Erziehungsrat* for permission to conclude the summer session early for reasons of poor health. During that year, he was forced to spend some time at a health resort in Tarasp in the mountains of Grisons. Repeatedly during the winter of 1915 and the spring of 1916, he found himself forced to turn over his inorganic lectures to Paul Pfeiffer. After Pfeiffer had left for Rostock in the winter semester of 1916, the lectures were given by Werner's assistant, Carl Agthe. Almost to the very end, Werner tenaciously refused to believe in the critical nature of the disease. His indomitable will-power is reflected in the many poignant letters which document the advancing illness.

18 April 1916 (*Regierungsrat*):

I do not yet feel able to hold this lecture [organic chemistry] since I still suffer from headaches for hours and days at a time. My physician, Dr Genhart writes . . . "It is my opinion that you should give up the entire summer semester [April–July, 1916]. If you will devote yourself entirely to regaining your health and eliminate all pressures you will recover completely."

Werner added that he hoped to resume full activity in the fall. Paul Pfeiffer, and later Carl Agthe, frequently substituted for him in the organic lectures.

5 February 1917 (*Erziehungsdirektion*):

Since I am still not yet recovered sufficiently to be able to resume the lectures without danger of a relapse, I find it necessary to ask the Erziehungsbehörde for a leave for the summer semester 1917 . . . . My physicians unanimously assure me that I shall recover and shall be able to resume my profession, since my present condition is the result of extreme nervousness . . . . [They recommend] complete relaxation and absolute rest . . . . For this purpose I wish to take up a lengthy residence in a sanatorium.

In accordance with his physicians' advice, Werner entered the Sanatorium La Charmille at Riehen, a suburb of Basel, while Prof. Haruthiun Abeljanz temporarily assumed the general supervision of *Abteilung* (Section) A.

A letter of 18 June 1917 to Carl Agthe is especially pathetic. The firm, precise pen strokes of the hale and sturdy Master have been replaced by the wavering script of a man reduced to a feeble shadow of his former self. Yet Werner optimistically insists that the worst is over and that his recovery is certain. 'The entire illness was exhaustion caused by overwork, and since the illness developed slowly, it will also retreat slowly.'

During the winter semester 1917/18, Werner made a valiant attempt to resume his lecture in inorganic chemistry and the directorship of the institute. This last return, heroic and tragic at the same time, could not help but be a time of intense suffering, for however hard Werner might have tried to suppress the knowledge, he must have realized that his mental faculties were rapidly deteriorating. He would forget the names, not to mention the research, of his *Doktoranden*. His lapses of memory were extremely erratic; in the mornings he would remember only events of the previous mornings, and in the afternoons, only those of the previous afternoons.

Difficulties in speech and articulation began to appear. He would rehearse each lecture word for word with his assistants, and even then he would at times be unable to complete the actual lecture. An apologetic assistant would have to finish the lecture of the professor who a few short years before had held his crowded audiences spellbound with his famed oratory!

Witnessing this gradually increasing incapacity was, of course, most painful to those who had known Werner at the height of his power – his *Doktoranden* of longer standing, his colleagues and his family. But many of the younger students, confronted only with the Werner of 1917–1918, reacted with the impatient intolerance of youth. Perhaps the cruelest blow of all came on 24 July 1918 when the Erziehungsrat received a petition with 42 signatures which began by declaring that the petitioners had 'the highest respect' for Werner as a scientist and scholar, but that 'in the interest of their academic careers' they were forced to lodge numerous complaints. The petition closed with an urgent plea for 'an improvement in a situation which has become intolerable'.

Werner's lectures during the summer semester of 1918 proved to be his last. On 28 August 1918 he wrote to the Regierungsrat:

> I am forced to request a complete leave for the winter semester 1918. I hope to return fully recovered next spring for the summer semester 1919 . . . . My nerves are not yet in order. Therefore I suffer from very quick exhaustion . . . . [My physi-

cians] strongly urge me to take off for a longer period of time and to rest completely if I wish to recover my former vigor and capacity for work . . . . I have been working for 25 years without a long vacation, and my body now demands a long and complete rest.

As a result of this letter, Professor Paul Karrer substituted for Werner both as head of the institute and as lecturer.

We may assume that Werner, struggling resolutely against his fate even on his deathbed, never relinquished his faith in his ultimate recovery. This hope remained with him until his illness had advanced so far that he was mentally incapable of himself writing his request to be retired at the end of the summer semester 1919. This sad duty fell to Frau Prof. Werner on 6 May 1919. Her letter to the *Erziehungsdirektion* was accompanied by a statement from Dr Genhart: 'The illness has unfortunately made such advances that a resumption of his activity at the university is unthinkable'. Werner's retirement became official on 15 October 1919.

Exactly one month later, on 15 November 1919, at Burghölzli, a Zürich psychiatric institution, death at last released Werner from his long physical and mental suffering. His body was cremated, and his ashes now rest in the family plot at the Rehalp Cemetery in Zürich. The funeral oration delivered at the grave on 17 November 1919 by Theodor Vetter, *Rektor* of the Universität Zürich, movingly recalls the cruel and heartbreaking metamorphosis:

> Anyone who met him a few months ago was painfully touched by the breakdown of his powers; anyone who had seen him in full activity received the impression of a victorious, inflexible, intellectual fighter for whom no task was too hard, no problem too difficult, before whom all obstacles had to give way. Let us remember this picture of him during this last hour . . . a man with many excellent qualities who conscientiously used his talents in the service of science and teaching. Let us remember this man here and not the frail invalid whom terrible suffering dragged slowly to his death.

The passage of more than half a century has not dimmed the magnitude of Werner's achievements. In fact, with the perspective of the passing years, we can better appreciate his monumental and revolutionary contributions. Regardless of what the future holds in store for chemistry, Alfred Werner will be remembered not only as the founder of modern inorganic stereochemistry but also as one of the major chemical figures of all time. In the next chapter we shall examine one of Werner's lectures that summarizes his achievements in his own words.

3

Molecularly asymmetric metallic compounds

(A Lecture by Alfred Werner Published in 1912)

Each volume in the 'Nobel Prize Topics in Chemistry' series contains an English version of an important work by the Nobel laureate whose work is featured in that particular book. In the case of Werner, a number of his most important works are already available in English translation. My earlier volume, *Classics in Coordination Chemistry, Part I* (Kauffman, 1968), contains annotated translations of six of his most important articles on coordination chemistry, each with introductory essays and commentary. Furthermore, Werner's Nobel address, 'Über die Konstitution und Konfiguration von Verbindungen höherer Ordnung' (Werner, 1913), has been translated into English by the Nobel Foundation. I have therefore chosen to include here an English translation of a French lecture 'Sur les composés métalliques à dissymétrie moléculaire', which Werner delivered before the Société Chimique de France at Paris on 24 May 1912. The lecture, which was published in the *Bulletin de la Société Chimique de France* [4] 11, No. 14, I–XXIV (20 July 1912), appeared in English translation in the *American Chemical Journal* 48, 314–336 (1912), which ceased publication in 1913 and hence is not readily available. The lecture summarizes succinctly Werner's most important work on the constitution and configuration of coordination compounds – work that we shall consider in more detail in Chapter 6. I have corrected typographical errors, added explanations where it seemed appropriate, but retained Werner's original nomenclature (Werner, 1897). What follows, then, is Werner's own description of the research which he carried out to prove his coordination theory, with special emphasis on optically active coordination compounds.

MOLECULARLY ASYMMETRIC METALLIC COMPOUNDS

Gentlemen, permit me, first, to extend my sincere thanks to your President for the honor which he has done me in inviting me to lay before you the results of our investigations on molecularly asymmetric metallic compounds. These studies may be considered as a continuation of the investigations begun by one of your most famous countrymen, by Pasteur, and I could not better begin my report than by expressing the great satisfaction I feel in having been able to widen, to a small extent, the field of the application of his great principle of *molecular asymmetry** and to continue the work of Le Bel and Van't Hoff.

Gentlemen, the hypotheses on the arrangement in space of the atoms of the molecules of carbon compounds [by Le Bel and Van't Hoff in 1874 — GBK] did not take a precise form until the tetravalence of carbon was clearly established [by August Kekulé in 1858 — GBK]. It was, in fact, only after this fundamental principle in the structure of organic molecules had been revealed and it had led to a general point of view and a classification of organic compounds, that it was possible to take up the question of the arrangement in space of the four groups which are combined with the carbon atom and to explain certain phenomena of isomerism by means of considerations on the different arrangement in space of these four groups.

The development of the theoretical conceptions relative to the stereochemical formulas of inorganic compounds followed an entirely analogous course. It was seen first that a large number of elementary metallic atoms have the property of forming complex radicals, MeA_6, in which the metallic atom Me is combined directly with the six groups A [Werner, 1893 — GBK], so that these complex radicals must correspond to the structural formula:

$$
\begin{array}{ccc}
A & & A \\
 \diagdown & & \diagup \\
A & \!\!\!- Me - \!\!\! & A \\
 \diagup & & \diagdown \\
A & & A
\end{array}
$$

The exactitude of this structural formula was established as the result of numerous investigations on the number of ions which the complex inorganic compounds form in aqueous solutions, investigations based especially on the determinations of the electrolytic conductivity of aqueous solutions of these compounds [Werner and Miolati, 1893, 1894 — GBK], for the groups directly combined with the central atom remain combined with the latter when the compounds are dissolved in water and consequently take no part in the electrolytic conductivity and do not appear in the form of ions free and independent of the central atom.

A new decisive experimental proof of the structural formula of these complex radicals was furnished by the discovery of a great number of new isomerism phenomena [for a discussion of Werner's research on the different types of structural isomerism see Kauffman, 1973c — GBK], predicted by the theory, as, for example, coördination polymerism, coördination isomerism, ionization metamerism, hydration isomerism, salt isomerism, isomerism phenomena, of which I had the honor to speak to you, some years ago, in an address made in Haller's laboratory [Werner, A. (1906), *Rev. gén. sci. pures et appliq.* **17**, 538 — GBK]. Thanks to these investigations on the constitution of complex inorganic compounds, the important conception of the coördination index [usually called coordination number – GBK] of elements was acquired, a conception which may now be summarized as follows: the coördination index of a large number of elementary atoms is equal to six, i.e. they have the power of combining with six other atoms.

We cannot take up here the nature of the forces which unite the six groups to the central atom; we shall note only that the affinity brought into play may manifest itself either in the form of principal valences or in the form of secondary valences.

As examples of compounds containing a complex radical, MeA_6, we may cite the following:

$$[Co(NO_2)_6]R_3 \quad [PtCl_6]R_2 \quad [AlF_6]Na_3 \quad [Fe(CN)_6]R_4 \quad [Co(NH_3)_6]X_3$$

$$\begin{bmatrix} (NH_3)_2 \\ Pt \\ Cl_4 \end{bmatrix} \quad \begin{bmatrix} NH_3 \\ Pt \\ Cl_4 \end{bmatrix}R \quad [Cr(OH_2)_6]X_3 \quad \begin{bmatrix} OH \\ Ru(NH_3)_4 \\ NO \end{bmatrix}X_2 \quad etc.$$

The number of compounds containing a radical of this type is very large.

Just as in the discussion of stereochemical problems relating to organic compounds the structural formula, CH_4, of methane serves as basis, so the structural formula, MeA_6, of the complex radicals serves as basis for the study of the configuration in space of inorganic molecules.

The first principal extension of our knowledge of the constitution of the complex radicals MeA_6 was gained experimentally and led to the admission that the six coördination positions of the central elementary atom are equivalent. This conception is derived from the impossibility of preparing isomeric compounds containing complex radicals, $Me_B^A{}_5$. We conclude that in the radicals MeA_6, the arrangement of the six groups around the central atom is symmetrical.

Theoretically, three different symmetrical arrangements can be imagined – the plane, the prismatic and the octahedral arrangement.

With the first two of these arrangements we can conceive of three series of isomers of compounds with the complex radicals $Me_{B_2}^{A_4}$; the octahedral arrangement, on the other hand, demands but two. Here again, it was experimental investigation which served to answer the question as to which of the arrangements mentioned corresponds to the configuration of these complex radicals. [See Chapter 6 for further details – GBK.] By means of numerous researches, it has been possible to show that there are never three series of isomers, but only two, and there are now known for the compounds of cobalt about 30 types of compounds with the complex radicals $Me_{B_2}^{A_4}$ which exist in the form of two series of stereochemical isomers; similar compounds of chromium and platinum are known. [For a detailed discussion of Werner's research on geometric isomerism, see Kauffman, 1975a – GBK.]

We can conclude with certainty that the six groups occupy the octahedral arrangement around the central atom. The problem of the determination of the configuration formulas of the isomers, i.e. of the determination of the relative positions occupied in the octahedral arrangement by the groups B of the isomeric radicals $Me_{B_2}^{A_4}$, has likewise been solved in a satisfactory manner [Werner, 1912 – GBK].

The configuration formulas deduced for the octahedral arrangement for the isomers with the complex radicals $\text{Me}_{B_2}^{A_4}$ contain the two groups B in different and opposite positions; in one of the forms they are in a near (*cis*) position and in the other in a distant (*trans*) position.

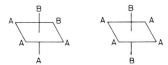

These stereochemical formulas show, therefore, differences analogous to those of the formulas of maleic and fumaric acids.

$$
\begin{bmatrix}
\begin{array}{c}
\text{HOOC}-\text{C}-\text{H} \\
\phantom{\text{HOOC}-}\|\phantom{-\text{H}} \\
\text{HOOC}-\text{C}-\text{H}
\end{array}
\quad \text{and} \quad
\begin{array}{c}
\text{HOOC}-\text{C}-\text{H} \\
\phantom{\text{HOOC}-}\|\phantom{-\text{H}} \\
\text{H}-\text{C}-\text{COOH, respectively} - \text{GBK}
\end{array}
\end{bmatrix}
$$

and we could therefore expect differences in properties between the inorganic stereochemical isomers analogous to those found between maleic and fumaric acids. It has been possible to verify experimentally these theoretical previsions. Of the two series of stereochemical isomers, only one shows direct relations with compounds containing in the place of the two groups B a radical forming a closed chain with the central atom, from which we may conclude that only in this series of isomers are the two groups B in a near (*cis*) position, in a position favorable for the closing of the chain.

With the experimental sanction of these relations and others, it has been possible to determine the configuration formulas of all the isomeric series, and the configurations thus established have been confirmed by the researches on optical isomers [for definitions of terms involved in optical isomerism, see Kauffman, 1972g, and for a discussion of these isomers, see Kauffman, 1974b — GBK], with which their study is related. From the octahedral formula, in fact, we can predict, besides the stereochemical isomers already mentioned, others which belong to the group of isomers with nonsuperposable images, and by the experimental confirmation of these extreme conclusions a new decisive proof for the octahedral formula was obtained, for these phenomena of optical isomerism can neither be foreseen nor explained by any other theoretical conception. Let us first take up briefly the theoretical conclusions deduced from the octahedral arrangement.

The octahedral arrangement leads, for a number of inorganic complex compounds, to configuration formulas with nonsuperposable images [i.e. their molecules should be asymmetric and therefore theoretically resolvable — GBK]. I should like, however, to limit my theoretical conclusions to simple cases easily attacked experimentally. Cases of this kind are found among compounds with the complex radicals $\text{Me}_{B_2}^{A_4}$, i.e. among the compounds which we have already taken up in our development of stereochemical isomerism. When the composition of the complex radicals $\text{Me}_{B_2}^{A_4}$ is such that 2 coördinately bivalent groups, such as ethylenediamine or other groups of analogous constitution, are substituted for the four groups A, and the two groups B are in the near (*cis*) position, the space formulas of these compounds are not superposable on their images. We may distinguish three different cases:

a. Compounds with complex radicals $\begin{bmatrix} A \\ MeEn_2 \\ B \end{bmatrix}$.

These complex radicals contain two asymmetric tetrahedrons (A, B, En, Me) which, as the result of the different orientation in space of the two molecules of ethylene-diamine (En), are not identical.

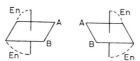

b. Compounds with complex radicals $\begin{bmatrix} B \\ MeEn_2 \\ B \end{bmatrix}$.

These radicals contain no asymmetric tetrahedral grouping. We therefore have there a kind of molecular asymmetry which we shall designate by the expression 'molecular asymmetry I'.

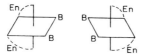

c. Compounds with complex radicals $[MeEn_3] X_3$.

These complex radicals contain three equal, coördinately bivalent groups. We shall call this asymmetry 'molecular asymmetry II'.

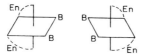

The stereochemical formulas of the *trans* isomers of all these compounds are superposable on their images and consequently we can expect no optical isomerism for these compounds.

The results of our researches harmonize perfectly with these theoretical deductions. We have been able to resolve into their optically active modifications the *cis* forms of the compounds of all the formula-types indicated above, but all efforts to resolve the *trans* isomers were unsuccessful. We may note, further, that in all cases the two active modifications have equal and opposite rotatory powers.

I shall point out now, in what follows, how these resolutions have been effected in practice and what the properties of the optical isomers are.

Methods of resolution

Thus far, we have not employed in our researches the spontaneous and biochemical methods of resolution; we have limited ourselves to the method generally employed in organic chemistry for the resolution of active bases. We caused the halogen salts of the series to be resolved to react with the silver salts of optically active acids, and then separated into their components, by means of fractional crystallization, the mixtures of salts formed. Among the acids which were used, we may mention especially the α-bromocamphorsulphonic acids and Reychler's camphorsulphonic acids. [For details on resolution see Chapter 6, pp. 121-135 — GBK.]

The α-bromocamphorsulphonic acids are much preferable to the camphorsulphonic acids, for thus far we have been able to resolve only two series of compounds by means of the camphorsulphonates, while we have resolved eight of them with the bromocamphorsulphonates.

Nevertheless, α-bromocamphorsulphonic acid is not a general agent for the resolution, for we have found that different series form partial racemates, so that the resolution is not successful. This, for example, is the case in the 1,2-diamminediethylene-diaminecobaltic and the carbonatodiethylenediaminecobaltic series [cis-[Co(en)$_2$-(NH$_3$)$_2$]$^{3+}$ and [Co(en)$_2$CO$_3$]$^+$, respectively — GBK] and others.

As regards the resolution by means of the d-α-bromocamphorsulphonic and d-camphorsulphonic salts, it should be noted that the isolation of the active series forming slightly soluble salts is easy, but the purification of the sometimes exceedingly soluble salts of the series of opposite rotatory power often presents difficulties. To get around these difficulties, we have, in certain cases, retransformed the soluble sulphonates into halogen salts and caused these to react with the l-bromocamphorsulphonate or camphorsulphonate of silver, so as to obtain difficultly soluble salts of the series of opposite rotatory power. By this method we have been able to isolate quantitatively the optical isomers.

In certain cases, the l-bromocamphorsulphonic acid, which is difficultly accessible, can be replaced by d-camphorsulphonic acid, for the solubility of the camphorsulphonates is often the inverse of that of the corresponding α-bromocamphorsulphonates. It is possible, therefore, in preparing the d-camphorsulphonate, to first isolate one of the series, then, starting with the salts remaining in the mother liquor, to prepare the d-camphorsulphonate of the opposite series. It goes without saying that the order of preparation of the bromocamphorsulphonates and camphorsulphonates can be reversed.

We likewise made numerous attempts to resolve the tartrates of different series by subjecting them to fractional crystallizations, but the results obtained were not satisfactory. On the other hand, we were able to determine that the resolution of compounds with trivalent radicals, corresponding to the general formula [MeEn$_3$] X$_3$, can be effected with remarkable success by means of mixed salts, such as the chloride-tartrates and bromide-tartrates:

$$\begin{array}{cc} \text{Cl} & \text{Br} \\ [\text{MeEn}_3] & \text{and} \quad [\text{MeEn}_3] \\ \text{C}_4\text{H}_4\text{O}_6 & \text{C}_4\text{H}_4\text{O}_6 \end{array}$$

which do not form partial racemates.

The method of fractional crystallization which we have just described has rendered us great service, but as it involves crystallizations in aqueous solutions it is applicable only to the case of compounds stable in aqueous solution. Therefore compounds with complex radicals containing groups which, by the action of water, are easily dissociated into the form of ions are almost excluded from this method. It is thus that the resolution of the 1,2-chloronitrodiethylenediaminecobaltic salts [cis-[Co(en)$_2$-(NO$_2$)Cl]$^+$ — GBK] was effected only with a very great loss of product.

To overcome these difficulties, we devised a method which is applicable to cases where the differences of solubility between the bromocamphorsulphonates or camphorsulphonates of the optical isomers are very large.

This method has, on the one hand, the advantage of not requiring the silver salts of the active acids, which, when the complex radicals contain a halogen, may give rise to secondary reactions, and furthermore avoids a long standing of the salts to be resolved in the aqueous solution, which is quite important in the cases where these salts are easily -altered by water. The new method is based on a simple precipitation process. If active ammonium bromocamphorsulphonate is added to a concentrated solution of the racemic compound to be resolved, one of whose active components gives a difficultly soluble bromocamphorsulphonate, the latter is precipitated, in some cases in pure state, in others mixed with a little of the partial racemate. (In the latter case, the precipitated salt is transformed into an easily soluble salt, and its solution is again precipitated with ammonium bromocamphorsulphonate.) From the mother liquors from the first precipitation, the series with opposite rotatory power can be isolated by adding the isomeric active ammonium bromocamphorsulphonate. By means of this method, it has been possible to resolve the cis-dichlorodiethylene-diaminecobaltic series [cis-[Co(en)$_2$Cl$_2$] X — GBK], which, in aqueous solution, is very rapidly transformed into the chloroaquacobaltic series [cis-[Co(en)$_2$(H$_2$O)Cl] X$_2$ — GBK]. But the best results have been obtained in the cis-chloroisosulphocyano-diethylenediaminecobaltic [modern, cis-chloroisothiocyanatobis(ethylenediamine)-cobalt(III), cis-[Co(en)$_2$(NCS)Cl] X — GBK] series, because the d-bromocamphor-sulphonate of the l-rotatory form and the l-bromocamphorsulphonate of the d-rotatory form of this series are almost insoluble in water. The importance of the new method has been increased lately by the use of the sodium salt of nitrocamphor, which we shall call sodium camphornitronate. This compound has made possible the resolution of series which could not be resolved by means of the bromocamphor-sulphonates and camphorsulphonates. The method of using sodium camphornitronate is the same as that for the ammonium salts of camphor- and bromocamphorsulphonic acids. Likewise, ammonium tartrate has rendered services in one case where the other methods did not succeed. We have found that, for the resolution of irontri-α-dipyridyl [[Fe((C$_5$H$_4$N)$_2$)$_3$] X$_3$ — GBK], ammonium tartrate can be used in the same way as the ammonium camphorsulphonate, ammonium bromocamphorsulphonate and sodium camphornitronate in the other cases.

Summarizing the new points of view bearing on the resolution of inorganic racemic compounds which result from our investigations, we reach the following conclusions:

1. Certain mixed salts, i.e. salts which, besides radicals of active acids, also contain radicals of inactive acids, such as the chloride-tartrates and bromide-tartrates, lend themselves in a peculiar manner to the resolution by fractional crystallization of racemic inorganic compounds.

2. Instead of the method of resolution generally used in organic chemistry, which is based on fractional crystallization and which is often inapplicable to inorganic compounds, advantageous use can be made of a new method, which consists in precipitating one of the active components from the aqueous solution of the racemate by means of the soluble salts of active acids.

General summary of active metallic compounds

Let us pass now to a general summary of the active metallic compounds thus far obtained. By fractional crystallization, the following series of metallic compounds have been resolved into their active modifications:

1. 1,2-Chloroamminediethylenediaminecobaltic series
   $[cis\text{-}[Co(en)_2(NH_3)Cl]X_2 - GBK]$.

2. 1,2-Bromoamminediethylenediaminecobaltic series
   $[cis\text{-}[Co(en)_2(NH_3)Br]X_2 - GBK]$.

3. Tetraethylenediamine-μ-aminoperoxodicobaltic series

   $[[(en)_2Co \begin{smallmatrix} \cdot O_2 \cdot \\ \cdot NH_2 \cdot \end{smallmatrix} Co(en)_2]X_4$. The Greek letter *mu* (μ) is used to designate

   groups bridging metal atoms $-$ GBK].

4. Tetraethylenediamine-μ-aminonitrodicobaltic series

   $[[(en)_2Co \begin{smallmatrix} \cdot NO_2 \cdot \\ \cdot NH_2 \cdot \end{smallmatrix} Co(en)_2]X_4 - GBK]$.

The resolution of these four series was effected with the bromocamphorsulphonates. The 1,2-dinitrodiethylenediaminecobaltic series $[cis\text{-}[Co(en)_2(NO_2)_2]X -$ GBK] was resolved by means of the bromocamphorsulphonates and the camphorsulphonates. For the triethylenediaminecobaltic and triethylenediaminerhodic series $[[Co(en)_3]X_3$ and $[Rh(en)_3]X_3$, respectively $-$ GBK], the resolution succeeded with the chloride-tartrates; for the triethylenediaminecobaltic series $[[Co(en)_3]X_3 -$ GBK], with the bromide-tartrates also.

By precipitation of concentrated solutions of racemates with ammonium *d*-bromocamphorsulphonate, we have been able to obtain the difficultly soluble bromocamphorsulphonates of the following series:

1. *d*-Chloroamminediethylenediaminecobaltic series
   $[d\text{-}cis\text{-}[Co(en)_2(NH_3)Cl]X_2 - GBK]$.

2. *d*-Bromoamminediethylenediaminecobaltic series
   $[d\text{-}cis\text{-}[Co(en)_2(NH_3)Br]X_2 - GBK]$.

3. 1,2-*l*-Dichlorodiethylenediaminecobaltic series
   $[l\text{-}cis\text{-}[Co(en)_2Cl_2]X - GBK]$.

4. *d*-Chloronitrodiethylenediaminecobaltic series
   $[d\text{-}cis\text{-}[Co(en)_2(NO_2)Cl]X - GBK]$.

5. *l*-Chloroisosulphocyanodiethylenediaminecobaltic series
   $[l\text{-}cis\text{-}[Co(en)_2(NCS)Cl]X - GBK]$.

6. l-Dichlorodiethylenediaminechromic series
$[l\text{-}cis\text{-}[Cr(en)_2 Cl_2] X - GBK]$.

With ammonium d-camphorsulphonate, we were able to precipitate the l-chloronitrodiethylenediaminecobaltic and l-bromonitrodiethylenediaminecobaltic camphorsulphonates $[l\text{-}cis\text{-}[Co(en)_2 (NO_2)Cl] C_{10} H_{15} OSO_3$ and $l\text{-}cis\text{-}[Co(en)_2 (NO_2)Br]\cdot$ $C_{10} H_{15} OSO_3$ − GBK]; finally, with ammonium d-tartrate, the tartrate of l-rotatory irontri-β-dipyridyl $[l\text{-}[Fe((C_5 H_4 N)_2)_3] C_4 H_4 O_6$ − GBK]; and with sodium camphornitronate, the d-triethylenediaminechromic and l-triethylenediaminerhodic camphornitronates $[d\text{-}[Cr(en)_3] (C_8 H_{14} CO\cdot C\cdot NO_2)_3$ and $l\text{-}[Rh(en)_3] (C_8 H_{14} CO\cdot C\cdot NO_2)_3$ − GBK].

Let us add to this enumeration the active metallic compounds prepared by chemical reactions from active compounds obtained by resolution. They are the following:

$$\begin{bmatrix} H_2O \\ \quad CoEn_2 \\ H_3N \end{bmatrix} X_3 \qquad \begin{bmatrix} H_3N \\ \quad CoEn_2 \\ H_3N \end{bmatrix} X_3 \qquad \begin{bmatrix} O\cdot \\ OC \quad CoEn_2 \\ O\cdot \end{bmatrix} X \qquad \begin{bmatrix} O_2N \\ \quad CoEn_2 \\ SCN \end{bmatrix} X$$

$$\begin{bmatrix} H_2O \\ \quad CoEn_2 \\ O_2N \end{bmatrix} X_2 \qquad \begin{bmatrix} \cdot NH_2\cdot \\ En_2Co \quad CoEn_2 \\ \cdot OH\cdot \end{bmatrix} X_4 \qquad \begin{bmatrix} \cdot NO_2\cdot \\ En_2Co \quad CoEn_2 \\ \cdot NH_2\cdot \end{bmatrix} X_4$$

By classifying all these series of compounds according to the radical types which characterize them, we obtain the following summary:

a. Compounds with the complex radicals $\begin{bmatrix} A \\ \; MeEn_2 \\ B \end{bmatrix}$ (asymmetric metallic atom):

(1) $\begin{bmatrix} Cl \\ \quad CoEn_2 \\ H_3N \end{bmatrix} X_2$ (2) $\begin{bmatrix} Br \\ \quad CoEn_2 \\ H_3N \end{bmatrix} X_2$ (3) $\begin{bmatrix} Cl \\ \quad CoEn_2 \\ O_2N \end{bmatrix} X$

(4) $\begin{bmatrix} Cl \\ \quad CoEn_2 \\ SCN \end{bmatrix} X$ (5) $\begin{bmatrix} O_2N \\ \quad CoEn_2 \\ SCN \end{bmatrix} X$ (6) $\begin{bmatrix} H_2O \\ \quad CoEn_2 \\ H_3N \end{bmatrix} X_3$

(7) $\begin{bmatrix} H_2O \\ \quad CoEn_2 \\ O_2N \end{bmatrix} X_2$ (8) $\begin{bmatrix} H_2O \\ \quad CoEn_2 \\ SCN \end{bmatrix} X_2$ (9) $\begin{bmatrix} \cdot O_2\cdot \\ En_2Co \quad CoEn_2 \\ \cdot NH_2\cdot \end{bmatrix} X_4$

(10) $\begin{bmatrix} \cdot OH\cdot \\ En_2Co \quad CoEn_2 \\ \cdot NH_2\cdot \end{bmatrix} X_4$ (11) $\begin{bmatrix} \cdot NO_2\cdot \\ En_2Co \quad CoEn_2 \\ \cdot NH_2\cdot \end{bmatrix} X_4$

b. Compounds with complex radicals $[A_2 MeEn_2]$ (molecular asymmetry I):

(1) $\begin{bmatrix} O_2N \\ \quad\quad CoEn_2 \\ O_2N \end{bmatrix} X$
(2) $\begin{bmatrix} Cl \\ \quad\quad CoEn_2 \\ Cl \end{bmatrix} X$
(3) $\begin{bmatrix} OC\begin{smallmatrix} O \\ \\ O \end{smallmatrix} CoEn_2 \end{bmatrix} X$

(4) $\begin{bmatrix} H_3N \\ \quad\quad CoEn_2 \\ H_3N \end{bmatrix} X_3$
(5) $\begin{bmatrix} Cl \\ \quad\quad CrEn_2 \\ Cl \end{bmatrix} X$

c. Compounds with complex radicals $[MeEn_3]$ (molecular asymmetry II):

(1) $[CoEn_3] X_3$ (2) $[CrEn_3] X_3$ (3) $[RhEn_3] X_3$ (4) $[Dip_3 Fe] X_2$

We thus know, in all, the two active forms of 20 series of metallic compounds, and these compounds are derived from four different elements: cobalt, chromium, iron and rhodium.

Characteristics of the active metallic compounds
After this general summary of the active compounds thus far obtained, we pass to the characterization of these compounds.

a. Cobalt compounds. The active compounds derived from cobalt which we have prepared may be classified in the following manner:

1. Diacidotetraminecobaltic series.
2. Acidopentaminecobaltic series.
3. Hexaminecobaltic series.
4. Series of active compounds with two asymmetric cobalt atoms.
5. Series of active compounds with asymmetric carbon and cobalt.

Of the diacidodiethylenediaminecobaltic salts, we may mention the following:

(1) $\begin{bmatrix} O_2N \\ \quad\quad CoEn_2 \\ O_2N \end{bmatrix} X$
(2) $\begin{bmatrix} O_2N \\ \quad\quad CoEn_2 \\ Cl \end{bmatrix} X$
(3) $\begin{bmatrix} SCN \\ \quad\quad CoEn_2 \\ Cl \end{bmatrix} X$

(4) $\begin{bmatrix} O_2N \\ \quad\quad CoEn_2 \\ Br \end{bmatrix} X$
(5) $\begin{bmatrix} Cl \\ \quad\quad CoEn_2 \\ Cl \end{bmatrix} X$
(6) $\begin{bmatrix} SCN \\ \quad\quad CoEn_2 \\ O_2N \end{bmatrix} X$

(7) $\begin{bmatrix} OC\begin{smallmatrix} O \\ \\ O \end{smallmatrix} CoEn_2 \end{bmatrix} X$

The stability of these compounds, dissolved in water, is very different. Thus, the dinitrodiethylenediaminecobaltic salts can stand for months in aqueous solution without their rotatory power* undergoing any change, while, on the other hand, the nitroisosulphocyanodiethylenediaminecobaltic salts are already less stable, for after four months we observed a decrease of rotatory power amounting to about one-fourth of the initial rotatory power.

The salts of the chloronitro series behave in a still different manner. The optical activity increases rapidly to a maximum value, which is about twice the first value; they show, therefore, the phenomenon of mutarotation*. It has been possible to determine the cause of this phenomenon; we have isolated the salts with increased rotation; they belong to the nitroaquodiethylenediaminecobaltic series:

$$\left[\begin{array}{l} O_2N \\ \qquad\quad CoEn_2 \\ OH_2 \end{array}\right] X_2$$

The phenomenon of mutarotation is therefore the result of a hydration. [The usual term is aquation* — GBK.] But the nitroaquodiethylenediaminecobaltic series undergoes, in aqueous solution, an ever-increasing racemization*, so that the rotatory power, after having attained a maximum, again diminishes, disappearing completely at the end of a certain time (2 days).

The dichloro series, $[Cl_2 CoEn_2] X$, behaves in an entirely different manner. The rotatory power, which is at first very large, diminishes rapidly and disappears completely after a few hours. We have here, therefore, a case of complete racemization taking place very rapidly. Likewise, the active salts of the carbonatodiethylenediaminecobaltic series behave in a very interesting manner. They are stable in cold aqueous solution, but if their solutions are heated the activity diminishes and it disappears completely if the temperature is raised to 90°. We have determined that the product formed is of the racemic carbonate series, and it is formed without its being possible, as in the other series, to observe the formation of intermediate aquocobaltic salts. This phenomenon of racemization completely resembles, therefore, the phenomena of autoracemization observed in organic chemistry. Let us pass now to an analysis of the chemical reactions of the active diacidocobaltic compounds. In the chloronitro compounds, we can substitute the chlorine by other acid radicals* without the production of any marked racemization. Thus, by the action of sodium nitrite are obtained the active dinitro salts and by the action of potassium sulphocyanate the active nitroisosulphocyanates:

$$\left[\begin{array}{l} Cl \\ \qquad\quad CoEn_2 \\ O_2N \end{array}\right] X \ + \ NaNO_2 \ = \ \left[\begin{array}{l} O_2N \\ \qquad\quad CoEn_2 \\ O_2N \end{array}\right] X \ + \ NaCl$$

$$\left[\begin{array}{l} Cl \\ \qquad\quad CoEn_2 \\ O_2N \end{array}\right] X \ + \ KCNS \ = \ \left[\begin{array}{l} SCN \\ \qquad\quad CoEn_2 \\ O_2N \end{array}\right] X \ + \ KCl$$

For the chloroisosulphocyanates, the reactions do not take place so sharply. Thus with sodium nitrite is obtained, besides the active nitroisosulphocyanate series, the racemic series also. The same thing occurs with the active dichloro salts which, with potassium carbonate, give much of the racemic carbonate together with a small quantity of the active salt. A complete racemization takes place when hydrochloric acid in alcoholic solution is made to react on a salt of the active carbonate series; completely inactive 1,2-dichloro salt is formed. In the same way the action of dilute mineral acids on the active carbonate salts produces only inactive diaquodiethylene-diaminecobaltic salts. All these facts show that, in certain reactions, intramolecular replacements in the asymmetric radicals take place very easily.

The active acidopentamine compounds which we have thus far prepared correspond to the formulas

$$\begin{bmatrix} Cl \\ \quad CoEn_2 \\ H_3N \end{bmatrix} X_2 \quad \text{and} \quad \begin{bmatrix} Br \\ \quad CoEn_2 \\ H_3N \end{bmatrix} X_2$$

These compounds are stable in cold or slightly warm aqueous solution; however, when they remain a long time in solution, they are partially transformed into aquo-amminediethylenediaminecobaltic salts, but without undergoing racemization. By the action of silver nitrate it is possible to transform the bromo series into the aquo-amminecobaltic series which we were able to isolate in the pure state. (Its salts show in 0.5 per cent solution a mean rotatory power $[\alpha]_D = 64°$ and $[M]_D = 392°$.) [The correct value of $[M]_D$ should be 256.1° $-$ GBK.]

If, on the other hand, the aquoamminecobaltic series is prepared by the action of an alkali on the bromo salts in concentrated solution, total racemization occurs. There seems to be here an important starting point for the experimental study of the problems of Walden's optical inversion.

The way in which the bromo series behaves when it reacts with liquid ammonia is very interesting. There are formed the two stereochemical isomers of the diammine-diethylenediaminecobaltic series, and while the *cis* form shows a rotatory power, the *trans* form is, as demanded by the theory, entirely optically inactive.

The group of active cobalt compounds with trivalent complex radicals obtained thus far consists of the following series:

$$[CoEn_3]X_3 \quad \begin{bmatrix} (NH_3)_2 \\ Co \\ En_2 \end{bmatrix} X_3 \quad \begin{bmatrix} HOH_2N \\ \quad CoEn_2 \\ H_3N \end{bmatrix} X_3 \quad \begin{bmatrix} Tn \\ Co \\ En_2 \end{bmatrix} X_3 \quad \begin{bmatrix} Pn \\ Co \\ En_2 \end{bmatrix} X_3$$

$$\begin{bmatrix} H_2O \\ \quad CoEn_2 \\ H_3N \end{bmatrix} X_3$$

[NH_2OH = hydroxylamine, Tn = triethylenediamine, Pn = propylenediamine $-$ GBK]

The triethylenediaminecobaltic series shows the strongest rotation (bromide, $[\alpha]_D = 117°$, $[M]_D = 600°$) [the correct value of $[M]_D$ is 560.4° $-$ GBK], and the amminehydroxylaminediethylenediaminecobaltic series has a rotatory power almost as large (bromide in 0.25 per cent solution, $[\alpha]_D = 112°$, $[M]_D = 545°$) [the correct value of $[M]_D$ is 525.2° $-$ GBK]. The diamminediethylenediaminecobaltic series shows the smallest rotatory power (bromide, $[\alpha]_D = 38°$, $[M]_D = 172°$).

Let us consider now the cobalt compounds with complex radicals containing two asymmetric cobalt atoms.

By causing ethylenediamine to react on the octammine-μ-aminodicobaltic salts are obtained tetraethylenediamine-μ-aminoperoxodicobaltic salts. The general formula for these salts is the following:

$$\left[\begin{array}{c} \text{III} \cdot \text{NH}_2 \cdot \text{IV} \\ \text{En}_2\text{Co} \qquad \text{CoEn}_2 \\ \cdot \text{O}_2 \cdot \end{array} \right] \text{X}_4$$

containing two asymmetric cobalt atoms. It has been possible to resolve this series by means of the bromocamphorsulphonates into two active forms with opposite rotatory powers.

These active forms have a rotatory power of exceptional magnitude; the bromides, for example, show a specific rotatory power* of 824°, whence is obtained a molecular rotatory power* of 6725°. [The correct value of [M] is 5981° — GBK.]

Still more interesting results have been obtained with the tetraethylenediamine-μ-aminonitrodicobaltic salts, which were prepared by the action of nitrous acid on the tetraethylenediamine-μ-aminoperoxodicobaltic series. They have a strong red-orange color. We have obtained three different d-bromocamphorsulphonates:

1. A difficultly soluble d-bromocamphorsulphonate derived from the d-μ-amino-nitro series.

2. A much more easily soluble d-bromocamphorsulphonate derived from the l-μ-aminonitro series [the original incorrectly reads 'd-aminonitro series' — GBK]. The salts prepared from these two bromocamphorsulphonates have equal but opposite rotatory powers.

3. A d-bromocamphorsulphonate of intermediate solubility, which gives inactive μ-aminonitro salts not resolvable into active forms.

By combining the active d- and l-salts, there are obtained racemic salts which are different from the inactive salts derived from the bromocamphorsulphonate of inter-mediate solubility.

We therefore have here a case analogous to that of the tartaric acids. We have, on the one hand, two active forms with opposite rotatory powers which, in equimolecular mixture, give a racemic form corresponding to 'racemic acid'. On the other hand, we have a series which is inactive by intramolecular compensation and corresponds to mesotartaric acid or 'nonresolvable tartaric acid'. [Werner thus not only proved that polynuclear as well as mononuclear complexes could be resolved but also demon-strated his theoretically predicted analogy between compounds containing two asym-metric carbon atoms and polynuclear complexes with two metal atoms, another striking confirmation of his octahedral hypothesis. In complete analogy with tartaric acid, HOOC·C*HOH·C*HOH·COOH (C* represents an asymmetric carbon atom), which, in addition to the racemic (d, l or ±) form, also exists in (+)- and (−)- forms and in an internally compensated nonresolvable (meso) form.

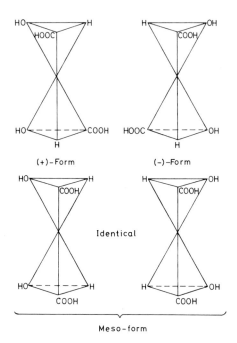

Werner was able to demonstrate experimentally for the brown binuclear complex

$$\left[(en)_2Co \underset{O_2N}{\overset{NH_2}{\diagdown \diagup}} Co(en)_2 \right] X_4$$

the existence of a racemic (±) form, (+)- and (−)-enantiomers, and an internally compensated, nonresolvable (meso) form

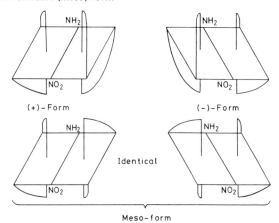

For further details of Pasteur's work see Kauffman, 1975f, 1977b — GBK.]

We may conclude from these facts that the constitutional formula of the tetra-ethylenediamine-μ-aminonitro series is symmetrical, i.e. that the nitro group is combined in the same way to the two cobalt atoms.

Let us note, likewise, that it is possible to transform the active salts into non-resolvable salts, a fact analogous to that observed for the tartaric acids by Jungfleisch, who, by heating tartaric acid at 200°, obtained mesotartaric acid. In the case we are considering, the transformation takes place under analogous conditions, for by vigorously boiling a solution of one of the active salts, a part of the active salt is transformed into a nonresolvable salt.

Let us conclude our summary of the results obtained with the active cobalt salts by adding that we have likewise prepared compounds containing at the same time a complex asymmetric radical and an asymmetric carbon, by the action of active propylenediamine on the dichlorodiethylenediaminecobaltic salts. As the propylenediamine enters the complex radical in the *cis* position, the two molecules of ethylenediamine which belong to the complex are perforce in the positions which condition the asymmetric arrangement of the complex radical. But as this arrangement leads to two stereochemical configurations, we should expect the formation of two isomeric active series, D-*l* and L-*d*, the salts of which should be distinguished by their optical activities [D and L refer to the configuration of the complex as a whole; *d* and *l* refer to the configuration of the propylenediamine — GBK].

These theoretical previsions have been confirmed experimentally. We obtained two series, which are distinguished especially by the different solubilities of their salts and their optical activity.

b. Chromium compounds. Thus far we have been able to prepare the optical isomers of two series of chromium compounds. They are the dichlorodiethylene-diaminechromic salts, which show the molecular asymmetry I, and the triethylene-diaminechromic salts, which exhibit the molecular asymmetry II. The dichloro-diethylenediaminechromic salts are very slightly stable in aqueous solution, like their analogues in the cobaltic series. Their molecular rotatory power, which is about 500°–600°, decreases very rapidly, and after a few hours the solutions have become entirely inactive. It is probable that this phenomenon of racemization is the result of the intermediate formation of hydrated [aqua — GBK] salts; however, we still lack a thorough experimental study of this subject.

The triethylenediaminechromic salts are, on the contrary, much more stable. Their molecular activity in cold aqueous solution is about 340° and hardly undergoes any change on long standing of the salts in solution. It is only when the solutions are evaporated that the optical activity diminishes [Werner never published any data on the resolution of [Cr(en)$_3$] X$_3$ compounds — GBK].

c. Iron compounds. All the active compounds of cobalt and chromium thus far obtained contain ethylenediamine as a constituent group of the complex molecule. It was therefore important to effect the resolution of compounds containing in the place of ethylenediamine other coordinately bivalent groups and to prove thus that the ethylenediamine is not absolutely necessary for the formation of active metallic compounds. Moreover, in order to be able to study the influence of the central atom on the rotatory power of complex inorganic compounds, it was desirable to prepare optically active compounds derived from other metallic elements. These considerations led us to attempt the resolution of the tri-α-dipyridylferrous compounds, [FeDpy₃]X₂. The discovery of these iron compounds is due to F. Blau, and they present a more general interest because they possess an intense red color and contain the iron in a masked* form, recalling the properties of hemoglobin. After numerous unsuccessful attempts, we have been able to prepare the tartrate of the *l*-rotatory series of irontri-α-dipyridyl by saturating the aqueous solution of the racemic bromide with ammonium *d*-tartrate. It was possible to transform the tartrate thus obtained into different salts, such as the iodide, the bromide, etc., and these salts were found to possess a very large rotatory power. The specific rotatory power varies around 500°, which, for the bromide, corresponds to a molecular rotatory power of 4000°.

In the solid state, these iron salts remain active indefinitely, but in aqueous solution they racemize very rapidly. After one-half hour, the rotatory power has already decreased by one-half, and after a few hours the solutions of these salts have become entirely inactive. This phenomenon of racemization is certainly caused by a partial decomposition of the active salts, with separation of a small quantity of α-dipyridyl, which, in aqueous solution, forms a dynamic equilibrium with the nondecomposed tri-α-dipyridylferrous salt. Our further studies must be applied to the investigation of active iron compounds showing greater stability. [Werner never published data on any other optically active iron complexes — GBK.]

The results obtained thus far with iron compounds show that the iron atom can likewise function as the center of asymmetric molecules showing optical activity. Moreover, they show that the activity of the optical isomers thus far described is not related with the specific nature of ethylenediamine but that the phenomenon also manifests itself when the ethylenediamine is replaced by other groups. Finally, let us emphasize the fact that in the iron compounds we have to do with derivatives of a bivalent metallic atom, while the active compounds of the other elements are derived from tri- and tetravalent metallic atoms.

d. Rhodium compounds. Although we do not, up to the present, know any rhodium compounds containing groups with two coordinative valences, we have been able to prepare the triethylenediaminerhodic salts. If sodium chlororhodiate, [RhCl₆]Na₃, is introduced into monohydrated ethylenediamine, it dissolves rapidly with evolution of heat and formation of a slightly yellow solution. This solution contains triethylenediaminerhodic chloride, [RhEn₃]Cl₃, which can be isolated in beautiful, colorless crystals. It has been possible to obtain the active triethylene-diaminerhodic salts by two different methods.

The first consists in precipitating the concentrated solution of the chloride with sodium camphornitronate. The camphornitronate which separates belongs to the *levo* series and by adding potassium iodide to the mother liquor the iodide of the *dextro* series is obtained.

The resolution is still better effected by the method used for the resolution of the triethylenediaminiecobaltic salts. Triethylenediaminerhodic chloride is made to react with a molecule of silver tartrate and the solution is concentrated until crystals are formed. The chloride-tartrate of the *levo* series separates in the form of colorless crystals, very clear and beautifully developed; the concentrated mother liquor gives a white salt of chalky appearance, which is the chloride-tartrate of the *dextro* series.

The chloride-tartrates can easily be transformed into other salts. The specific rotatory power of these salts varies between 70° and 80°, which corresponds to a molecular rotatory power of about 300°. [The correct values of [M] lie between 326.7° and 373.4° – GBK.]

Relations between the optical activity, the configuration, and the constitution of the complex inorganic compounds

The greater number of the active metallic compounds which we have studied contain a complex radical [MeEn$_2$], whose formula in space is not superposable on its image and can thus present itself under two stereochemical forms.

We might inquire if all the compounds presenting the same configuration of the radical [MeEn$_2$] deviate the plane of polarized light in the same direction. We were easily able to establish that this is not the case, but that the direction of the rotatory power depends on the nature of the radicals which occupy the two other coordination positions of the central atom and on the nature of that central atom.

This is derived from the fact that we have been able to prepare compounds of opposite rotatory powers by chemical reactions which are certainly not accompanied by a change in configuration. Let us give some examples: starting from the *l*-dichloro-diethylenediaminecobaltic series, we have obtained, by making it react with potassium carbonate, the *d*-carbonatodiethylenediaminecobaltic series:

$$\begin{bmatrix} Cl \\ \quad CoEn_2 \\ Cl \end{bmatrix} Cl + K_2CO_3 \longrightarrow 2KCl + \begin{bmatrix} O \\ OC \diagdown \quad CoEn_2 \\ O \end{bmatrix} Cl$$

$$Levo \qquad\qquad\qquad\qquad Dextro$$

Likewise, the *l*-chloroisosulphocyanodiethylenediaminecobaltic salts are transformed, by the action of sodium nitrite, into *d*-nitroisosulphocyanodiethylenediaminecobaltic salts:

$$\begin{bmatrix} Cl \\ \quad CoEn_2 \\ SCN \end{bmatrix} X + NaNO_2 \longrightarrow NaCl + \begin{bmatrix} O_2N \\ \quad CoEn_2 \\ SCN \end{bmatrix} X$$

$$Levo \qquad\qquad\qquad\qquad Dextro$$

Moreover, by reduction of the salts of the *l*-tetraethylenediamine-μ-aminoperoxodi-cobaltic series are obtained the salts of the *d*-tetraethylenediamine-μ-amino-oldi-cobaltic series [the term *ol* designates a hydroxyl group bridging two metal atoms – GBK]:

$$\begin{bmatrix} \cdot O_2 \cdot \\ En_2Co \qquad CoEn_2 \\ \cdot NH_2 \cdot \end{bmatrix} X_4 \longrightarrow \begin{bmatrix} \cdot OH \cdot \\ En_2Co \qquad CoEn_2 \\ \cdot NH_2 \cdot \end{bmatrix} X_4$$

$$Levo \qquad\qquad\qquad\qquad Dextro$$

Finally, by the action of nitrous acid on the tetraethylenediamine-μ-aminoperoxo-dicobaltic salts is obtained, starting from the l-series, the d-series of tetraethylene-diamine-μ-aminonitrodicobaltic salts:

$$\begin{bmatrix} & \cdot O_2 \cdot & \\ En_2Co & & CoEn_2 \\ & \cdot NH_2 \cdot & \end{bmatrix} X_4 \longrightarrow \begin{bmatrix} & \cdot NO_2 \cdot & \\ En_2Co & & CoEn_2 \\ & \cdot NH_2 \cdot & \end{bmatrix} X_4$$

$$\qquad\qquad Levo \qquad\qquad\qquad\qquad\qquad Dextro$$

On the other hand, we have been able to establish the following interesting law: the active isomeric series which give, with the same active acids, the less soluble salts show the same stereochemical configuration of the radical [MeEn$_2$]. This law is demonstrated by the following facts:

The d-chloronitro series and the l-chloroisosulphocyano series give with d-bromo-camphorsulphonic acid salts which are much less soluble than those formed with the l-acid. It is proved that they correspond to the same stereochemical configuration by the fact that, by the action of potassium sulphocyanide on a salt of the d-chloronitro series and of sodium nitrite on a salt of the l-isosulphocyano series, the same d-rotatory nitroisosulphocyano series is obtained:

$$\begin{bmatrix} Cl & \\ & CoEn_2 \\ O_2N & \end{bmatrix} X + KSCN \longrightarrow KCl + \begin{bmatrix} SCN & \\ & CoEn_2 \\ O_2N & \end{bmatrix} X$$

$$\begin{bmatrix} Cl & \\ & CoEn_2 \\ SCN & \end{bmatrix} X + NaNO_2 \longrightarrow NaCl + \begin{bmatrix} O_2N & \\ & CoEn_2 \\ SCN & \end{bmatrix} X$$

[The original reads KNO$_2$ and KCl — GBK.]

These transformations may be represented by the following scheme (giving difficultly soluble d-bromocamphorsulphonates):

$$\begin{bmatrix} Cl & \\ & CoEn_2 \\ O_2N & \end{bmatrix} X$$

$$Dextro$$

$$\begin{bmatrix} Cl & \\ & CoEn_2 \\ SCN & \end{bmatrix} X$$

$$Levo$$

$$\begin{bmatrix} O_2N & \\ & CoEn_2 \\ SCN & \end{bmatrix} X$$

$$Dextro$$

Another fact of the same nature is the following: the l-tetraethylenediamine-μ-aminoperoxodicobaltic series furnishes a difficultly soluble d-bromocamphor-sulphonate, while the d-tetraethylenediamine-μ-aminonitrodicobaltic series gives a difficultly soluble d-bromocamphorsulphonate.

Now, the l-tetraethylenediamine-μ-aminoperoxodicobaltic series is transformed by the action of nitrous acid into the d-tetraethylenediamine-μ-aminonitrodicobaltic series, proving that the configuration of the [CoEn$_2$] groups of these series is identical.

We may summarize the phenomena observed in the following manner. The compounds of the series with identical asymmetry show parallel processes of solubility in their salts with active acids, but the direction of the rotatory power of the salts in question is not necessarily the same.

It is on the basis of these new ideas that we can study in more detail the relations between the rotatory power and the composition of the asymmetric molecules.

Classifying the series with identical configuration in the order of the magnitude and direction of their molecular rotatory powers, we obtain the following summary:

$$\left[En_2Co\begin{matrix}\cdot OH\cdot \\ \\ \cdot NH_2\cdot\end{matrix}CoEn_2\right]X_4 \quad [Cl_2CoEn_2]X \quad [Cl_2CrEn_2]X \quad [RhEn_3]X_3$$

$$-6723° \qquad\qquad -550° \qquad\quad -400° \qquad\quad -350°$$

$$\left[\begin{matrix}Cl \\ \\ SCN\end{matrix}CoEn_2\right]X \quad \left[\begin{matrix}Cl \\ \\ O_2N\end{matrix}CoEn_2\right]X \quad \left[\begin{matrix}H_3N \\ \\ H_2O\end{matrix}CoEn_2\right]X_3$$

$$-200° \qquad\qquad +74° \qquad\qquad +124°$$

$$\left[\begin{matrix}O_2N \\ \\ O_2N\end{matrix}CoEn_2\right]X \quad \left[\begin{matrix}O_2N \\ \\ H_2O\end{matrix}CoEn_2\right]X_2 \quad \left[\begin{matrix}H_3N \\ \\ H_3N\end{matrix}CoEn_2\right]X_3$$

$$+145° \qquad\qquad +165° \qquad\qquad +172°$$

$$\left[\begin{matrix}H_3N \\ \\ Cl\end{matrix}CoEn_2\right]X_2 \quad \left[\begin{matrix}H_3N \\ \\ Br\end{matrix}CoEn_2\right]X_2 \quad \left[\begin{matrix}O_2N \\ \\ SCN\end{matrix}CoEn_2\right]X$$

$$+172° \qquad\qquad +180° \qquad\qquad +200°$$

$$\left[OC\begin{matrix}O \\ \diagup\diagdown \\ O\end{matrix}CoEn_2\right]X \quad [CrEn_3]X_3 \quad [CoEn_3]X_3$$

$$+280° \qquad +340° \text{ to } +370° \quad +600°$$

$$\left[En_2Co\begin{matrix}\cdot OH\cdot \\ \\ \cdot NH_2\cdot\end{matrix}CoEn_2\right]X_4 \quad \left[En_2Co\begin{matrix}\cdot NO_2\cdot \\ \\ \cdot NH_2\cdot\end{matrix}CoEn_2\right]X_4$$

$$+990° \qquad\qquad\qquad +1300° \text{ to } +1400°$$

As a result, the rotatory power is seen to vary between very wide limits, and all the groups forming part of the asymmetric complex radical have an influence on the value of the rotatory power. It is seen, likewise, that the nature of the central atom has a capital influence on the magnitude and the direction of the rotatory power, for the rotatory powers of the triethylenediaminecobaltic, -chromic, and -rhodic salts, as well as those of the dichlorodiethylenediaminecobaltic and -chromic salts, are entirely different. The compounds of chromium have a smaller rotatory power than the cobalt compounds, but this difference is not equal in the different series; it is about 150° for the dichloro salts, while it is about 250° for the salts containing the triethylenediamine complex radical. The magnitude of the rotatory power depends, therefore, on a special factor, the value of which is determined by the particular relations which exist between the central atom and the groups which it unites in the complex radicals.

Let us add that the triethylenediaminerhodic salts deviate the plane of polarized light almost as much to the left as the triethylenediaminechromic salts of the same configuration deviate it to the right.

The values for the rotatory powers which served for our theoretical deductions were determined under experimental conditions as nearly comparable as possible; nevertheless, these values probably do not yet give an exact measure of the relative magnitudes of these rotatory powers, and this for the following reasons: The magnitude of the rotatory power often depends to a very high degree on the concentration of the solutions of the salts, and, as nearly as we can judge thus far, the activity is greater the more dilute the solution. This is probably because the compounds in the undissociated state have a smaller rotatory power than their asymmetric cations, and it will be necessary, therefore, in order to obtain comparable values, to determine the rotatory power of the undissociated compounds and of their cations. Moreover, it is to be noted that almost all the compounds studied thus far are colored and show very large rotatory dispersions*, so that it will be necessary to make exact studies on the relations which exist between the magnitude of the rotatory power and the wave length of the light.

I hope that the researches which we have begun in this direction will lead to precise ideas on the influence of the different atomic groups of the complex radicals on the optical activity.

To end our review, let us consider briefly the consequences of the results thus far obtained for the theory of valence and for stereochemistry.

Our experiments show in the first place that it is of secondary importance for the stability of the molecules of the complex compounds whether the atoms are united by principal valences* or by secondary valences*. For the study of the structure of inorganic compounds, secondary valences assume an equal importance with principal valences. This naturally leads us to admit that there is no essential difference between the affinities brought into play by the principal valences and the secondary valences. We probably have to do with affinity forces which are distinguished only in magnitude and not in quality and the difference between which does not manifest itself in any notable manner except when the atomic unions which they produce are so different that the ones lead to compounds of great stability and the others to compounds which are slightly stable.

The question of knowing what secondary causes (migrations of electrons, etc.) may lead to a more pronounced differentiation between the atomic unions produced by principal and secondary valences as independent of the question of the nature of the affinity forces which determine the structure of complex inorganic molecules.

As to what concerns stereochemistry, our studies have led us to the probable conclusion that all elementary atoms, in so far as they can function as the central atom of stable complex radicals, are capable of forming compounds showing optical isomerism. Moreover, it has been shown that the optical isomerism is not necessarily due to the different nature of the groups combined with the central atom, but that every asymmetric configuration leads to optical isomerism, even if the constituent parts of the molecular grouping are equal.

We come back then, to the principle which Pasteur first formulated and which admits that every molecule having no plane of symmetry, i.e., showing structural asymmetry, must always exist in the form of two oppositely active optical modifications. The isomerism phenomena related with the molecular asymmetry of metallic compounds, with asymmetric cobalt, with asymmetric carbon, with asymmetric nitrogen, with the molecular asymmetry of the inositols, etc., are therefore only special cases resulting from this general principle, a principle from which the possibility of numerous other cases of optical isomerism can still be predicted.

The end of future investigations, then, will be to determine what other new cases of optical isomerism, foreseen through our conceptions on the arrangement in space of inorganic molecules, can be realized experimentally.

4

Some historically significant coordination compounds

Who discovered the first coordination compound? And when? Different historians and chemists give different answers to these questions. In this chapter we shall examine some historically significant coordination compounds (Kauffman, 1974e, 1977c).

ALIZARIN DYE

Perhaps the earliest known of all coordination compounds is the bright red alizarin dye, a calcium aluminum chelate compound of hydroxyanthraquinone:

Clay was the source of the calcium and aluminum ions, and the hydroxyanthraquinone was obtained from the roots of the madder plant found in Europe, Asia Minor and the British Isles. Although the exact origin of this dye is shrouded in antiquity, it was first used in India and was known to the ancient Persians and Egyptians long before it was used by the Greeks and Romans. Joseph's 'coat of many colors' mentioned in Chapter 37 of *Genesis* may possibly have been treated with alizarin, which is an excellent textile dye.

Alizarin is also mentioned by the Greek Herodotus, the father of history, in about 450 BC, and it was probably the red dye used by Alexander the Great to win a decisive battle against a much larger Persian army 120 years later. In what was probably the first recorded example of chemical warfare or camouflage, Alexander dyed the clothing of most of his soldiers with bloodlike splotches and enticed the Persians into heedlessly attacking what they thought was a demoralized force of badly wounded men. In more recent times, madder dyes were an integral part of American Revolutionary history, being the dye used in the British 'redcoats' (Martin and Martin, 1964).

TETRAAMMINECOPPER(II) ION

Probably the first scientifically recorded observation of a completely inorganic coordination compound is the formation of the familiar tetraamminecopper(II) ion, $[Cu(NH_3)_4]^{2+}$:

$$\begin{bmatrix} H_3N & & NH_3 \\ & Cu & \\ H_3N & & NH_3 \end{bmatrix}^{2+}$$

This observation is found in the writings of the sixteenth-century German physician and alchemist Andreas Libavius (1540?–1616), who noticed that *aqua calcis* (lime water, saturated calcium hydroxide solution, $Ca(OH)_2$) containing *sal ammoniac* (ammonium chloride, NH_4Cl) became blue in contact with brass (an alloy of copper and zinc) (Libavius, 1597).

PRUSSIAN BLUE

Another candidate for the first coordination compound is Prussian blue, potassium iron(III) hexacyanoferrate(II), a complex of empirical formula $KCN \cdot Fe(CN)_2 \cdot Fe(CN)_3$ with the structure:

$$K^+ \quad Fe^{3+} \begin{bmatrix} & N & \\ & C & \\ NC & | & CN \\ & Fe & \\ NC & | & CN \\ & C & \\ & N & \end{bmatrix}^{4-}$$

This substance was first obtained accidentally in 1704 by Diesbach, a manufacturer of artist's colors from Berlin, who caused this dark blue substance to precipitate from an iron-containing solution by adding an alkali obtained from the wandering German alchemist Johann Conrad Dippel. It was first described (Anon., 1710) as a nontoxic pigment suitable for oil colors, but its method of preparation was kept secret, probably because Diesbach wished to benefit monetarily from his discovery.

A recipe for the preparation of Prussian blue was published 20 years after its discovery (Woodward, 1724). The complicated process described as 'Preparation of Prussian Blue sent from Germany to John Woodward, MD Prof. Med. Gresh, FRS' is reproduced in the original Latin, accompanied by an English translation, in a short paper entitled 'The Beginnings of Co-ordination Chemistry' (Powell, 1959). The directions involve the use of crude tartar (potassium hydrogentartrate, $KHC_4H_4O_6$), dried crude nitre (potassium nitrate, KNO_3), charcoal, well-dried and finely powdered ox blood, green vitriol (ferrous sulfate or copperas, $FeSO_4$) calcined gently to whiteness, crude alum (potassium aluminum sulfate, $K_2SO_4 \cdot Al_2(SO_4)_3 \cdot 24H_2O$), and spirit of common salt (hydrochloric acid, HCl). Woodward cautions:

In this procedure the calcination is of great importance because the sea-blue colour and the hidden sky-blue arise according as the calcination of the dried blood with the alkali is light, medium, or strong, and hence there is a diversity of colour. The well-boiling lixivia [filtrates obtained by leaching soluble from insoluble matter, in this process by filtrations through linen] are to be mixed with the other in the most rapid manner.

John Browne showed that the alum is unnecessary and that the blue color cannot be obtained with metals other than iron (Browne, 1724). His conclusion that the origin of the color lies in the iron was the first step in the elucidation of the chemical constitution of Prussian blue, which was subsequently investigated by various chemists, who devised alternative methods of preparation. Since materials containing iron, potash and nitrogenous matter such as blood or animal hooves may have been heated together in more remote times, ferrocyanides (hexacyanoferrates(II)) probably antedate Diesbach's discovery. As the structure shows, the compound contains Fe—CN bonds and may thus possibly be considered as the first known example of a coordination compound containing transition metal—carbon bonds. In other words, it may be the first representative of the organometallic compounds whose chemistry is currently the subject of much intensive research. [See FUNDAMENTALS: *Organometallic Compounds*, in this series.]

HEXAAMMINECOBALT(III) ION

Most authorities attribute the discovery of the first metal—ammine to Tassaert, a Parisian chemist about whom virtually nothing is known — not even his first name. In his short article (Tassaert, 1798), he is identified only as Citoyen Tassaert — Citizen Tassaert — and even the Bibliothèque of the École Nationale Supérieure des Mines at Paris, where he is said to have worked, was unable to give me any information about him. Some chemists imply or even openly state that Tassaert was the first to prepare hexaamminecobalt(III) chloride, $[Co(NH_3)_6]Cl_3$:

$$\left[\begin{array}{c} NH_3 \\ H_3N \diagdown \quad \diagup NH_3 \\ Co \\ H_3N \diagup \quad \diagdown NH_3 \\ NH_3 \end{array} \right] Cl_3$$

the parent compound from which all cobalt–ammines may be considered to be derived. Yet he merely observed the brownish mahogany color of the solution formed when excess aqueous ammonia is added to a solution of cobalt chloride or cobalt nitrate, and he failed to follow up his accidental discovery.

Tassaert's article is entitled 'Analyse du Cobalt de Tunaberg, suivie de plusieurs moyens d'obtenir ce métal à l'état de pureté, et de quelques-unes de ses propriétés les plus remarquables' and deals with the dissolution and analysis of a cobalt ore from Tunaberg (Kolmården, Södermanland, Sweden). Tassaert used excess ammonia to dissolve the cobalt and simultaneously to precipitate the iron in the ore. The passage of interest to coordination chemists occurs on pp. 106–107:

> Another rather surprising phenomenon is that when nitrate of cobalt is precipitated by excess ammonia, there is formed a precipitate which immediately redissolves and gives a brown solution; but if this solution is diluted at once with a large amount of water, there is formed a green precipitate which consists only of pure oxide of cobalt, which dissolves in acids and imparts a beautiful pink color to the solutions; if on the contrary this solution of cobalt in ammonia is left exposed to the air for a long time, it can be diluted with as much water as one wishes without the formation of any precipitate. This fact, which I do not yet claim to explain, but which I intend to reconsider, has appeared to me to be worthy of recording here.

VAUQUELIN'S SALT AND MAGNUS' GREEN SALT
(Kauffman, 1974d, 1975f)

As already mentioned in our discussion of nomenclature, one of the ways of naming these complex compounds was after their discoverers.

Louis-Nicolas Vauquelin (1763–1829), Professor of Chemistry at the Collège de France, discovered the pink compound tetraamminepalladium(II) tetrachloropalladate, $[Pd(NH_3)_4][PdCl_4]$ (Vauquelin, 1813):

$$\begin{bmatrix} H_3N & & NH_3 \\ & Pd & \\ H_3N & & NH_3 \end{bmatrix}^{2+} \begin{bmatrix} Cl & & Cl \\ & Pd & \\ Cl & & Cl \end{bmatrix}^{2-}$$

This compound, which contains coordinated palladium in both the cation and the anion, is still known as Vauquelin's Salt after its discoverer. The corresponding platinum compound, $[Pt(NH_3)_4][PtCl_4]$:

$$\begin{bmatrix} H_3N & & NH_3 \\ & Pt & \\ H_3N & & NH_3 \end{bmatrix}^{2+} \begin{bmatrix} Cl & & Cl \\ & Pt & \\ Cl & & Cl \end{bmatrix}^{2-}$$

was discovered by Heinrich Gustav Magnus (1802–1870), Professor of Physics and Technology at the University of Berlin (Plate 4) (Magnus, 1828). It constitutes the first discovered platinum-ammine and is still known as Magnus' Green Salt. Both Vauquelin's Salt and Magnus' Green Salt differ markedly in color from their constituent ions, a fact which indicates interaction between the cationic metal atom and the anionic metal atom. The two salts, incidentally, are also so-called 'polymerization isomers' of the *cis* and *trans* isomers of the dichlorodiammine complexes

Plate 4.   Heinrich Gustav Magnus (1802–1870) [Prandtl, W. (1956), *Deutsche Chemiker in der ersten Hälfte des neunzehnten Jahrhunderts*, p. 303, Verlag Chemie, Weinheim/Bergstrasse]

of the corresponding metals, which for platinum are *cis-* and *trans-* [Pt(NH$_3$)$_2$Cl$_2$] (see p. 64). In other words, the salts have the same empirical formulae as the neutral dichlorodiammine compounds, but their formula weights are a multiple, namely two, of those of the dichlorodiammines (Kauffman, 1975f, 1976b).

GMELIN'S COMPOUNDS

According to some authorities, the first metal–ammine to be isolated in the solid state was the reddish yellow hexaamminecobalt(III) oxalate [Co(NH$_3$)$_6$]$_2$(C$_2$O$_4$)$_3$:

described by Leopold Gmelin (1788–1853), Professor of Medicine and Chemistry at the University of Heidelberg (Plate 5) (Gmelin, 1822a). In

Plate 5. Leopold Gmelin (1788–1853) [Partington, J. R. (1964), *A History of Chemistry*, Vol. 4, p. 181, Macmillan, London]

ine same year Gmelin also discovered several new double salts –
potassium ferricyanide, or potassium hexacyanoferrate(III), also known as
red prussiate of potash, K_3 [Fe(CN)$_6$] (Gmelin, 1822b):

the cobalticyanides or hexacyanocobaltates(III), M_3 [Co(CN)$_6$] (Gmelin,
1822c):

and the platinocyanides or tetracyanoplatinates(II), M_2 [Pt(CN)$_4$]
(Gmelin, 1822d):

All these substances certainly deserve to rank among the earliest known of
coordination compounds.

ZEISE'S SALT

An extremely interesting compound which played an important role in
the development of bonding theory in both inorganic and organic chemistry is potassium trichloro(ethylene)platinate(II) monohydrate:

which was discovered by William Christoffer Zeise (1789–1847), Professor of Chemistry at the University of Copenhagen (Plate 6) (Zeise,

Plate 6. William Christoffer Zeise (1789–1847) [(1919), *J. Prakt. Chem.* **99**, 281]

1827). The compound was the first discovered organometallic compound to contain an unsaturated organic ligand and is still known as Zeise's Salt. Its discovery led to the preparation and characterization of many other platinum–olefin compounds. Its structure and bonding, however, were not understood until the preparation and characterization of ferrocene (bis(cyclopentadienyl)iron(II)):

by Kealy and Pauson (1951) in the United States and by Miller *et al.* (1952) in England. This compound constitutes the first representative of the so-called 'sandwich' compounds that have created such a stir in inorganic circles in recent years (Kauffman, 1976b, 1979d).

PEYRONE'S SALT AND REISET'S SECOND CHLORIDE

In addition to Magnus' Green Salt, two other extremely important platinum(II) compounds were discovered during the first half of the nineteenth century. Both compounds possess the same formula – $Pt(NH_3)_2Cl_2$ – and both were discovered in the same year – 1844. They differ, however, in physical and chemical properties, and they constitute the simplest, best known, and longest known case of stereoisomerism among platinum compounds. One isomer, called platosemidiammine chloride or Peyrone's Salt, was first prepared by the action of aqueous ammonia on potassium tetrachloroplatinate(II) (Peyrone, 1844), while the other isomer, called platosammine chloride or Reiset's Second Chloride, was first prepared by the French chemist and agronomist Jules Reiset (1818–1896) by the action of heat or concentrated hydrochloric acid on tetraammineplatinum(II) chloride (Reiset, 1844). Werner, in his first paper on the coordination theory (Werner, 1893), discussed these compounds at length and considered them to be geometric isomers with a square planar configuration:

cis   trans

Platosemidiammine chloride   Platosammine chloride
Peyrone's Salt   Reiset's Second Chloride

a configuration that has since been amply corroborated. The explanation of the course of the reactions involved in the preparation of these isomers was finally given in generalized form by the Russian chemist Il'ya Il'ich Chernyaev (1893–1966) in his famous *trans* effect (see pp. 158–162).

GIBBS AND GENTH'S RESEARCHES (Kauffman, 1977e)

In addition to Gmelin's work, there were a few other investigations of cobalt–ammonia compounds during the first half of the nineteenth century, but they were isolated instances and dealt mostly with species in solution. Credit for the first distinct recognition of the existence of perfectly well-defined and crystallized solid salts of cobalt–ammines belongs to the now almost forgotten German–American chemist and mineralogist Frederick Augustus Genth (1820–1893) (Plate 7)(Kauffman, 1972f, 1975c).

It was in 1847, while he was Robert Bunsen's assistant at the University of Marburg and during Bunsen's absence in Iceland, that the 27-year-old

Plate 7. Frederick Augustus Genth (1820–1893) [(1902), *Nat. Acad. Sci.,* *Biog. Mem.* **4**, 202]

Genth obtained his first results on the cobalt–ammines. Their discovery, like many discoveries in science, was accidental, but as Louis Pasteur has said, 'Chance favors the prepared mind', and Genth was quick to recognize the import of the accident. The story of the discovery was passed by oral tradition from Genth to Edgar Fahs Smith to Thomas P. McCutcheon to Louis C. W. Baker. Drs McCutcheon and Baker were my instructors in general chemistry at the University of Pennsylvania. According to the story, Genth was demonstrating qualitative analysis to his students. Immediately before vacation, possibly Christmas, he had just removed the precipitated metals of Analytical Group II by filtration from acidic solution. The procedure called for making the solution basic with potassium hydroxide before resaturating with hydrogen sulfide. Since the demonstration laboratory had run out of potassium hydroxide, Genth substituted ammonia water, but he had no time to resaturate the basic solution with hydrogen sulfide, so he put it aside. After vacation, when he returned to precipitate the metals of Analytical Group III, he found large, beautiful, colored, inexplicable crystals of a sort not encountered before. Careful repetitions of the procedure, with exposure of the ammonia-containing cobalt solutions to air, but with varying conditions, enabled Genth to prepare several different types of crystals. He freely communicated his results to others and deposited samples of the salts in the laboratory at the University of Giessen.

Before Genth was able to complete his research on and analysis of these new compounds, he emigrated to the United States. Thus it was not until January of 1851 that he published his results as a short 4 1/2-page preliminary note in German in an obscure Philadelphia journal for German physicians (Genth, 1851). Here he described salts of two cations:

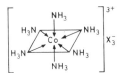

Aquapentaamminecobalt(III)   Hexaamminecobalt(III)
or roseocobalt   or luteocobalt

Of course, as we have already seen, a few similar compounds of platinum had been prepared in Europe before Genth, but his work drew attention to the neglected field of coordination compounds. In July of 1852, Oliver Wolcott Gibbs (1822–1908), an American chemist two years younger than Genth (Plate 8) (Kauffman, 1972b), began to collaborate with Genth on an investigation which has since become famous in the annals of coordination chemistry.

Several months later, in November of 1852, Gibbs, then at the Free Academy which later became the City University of New York, made his

Plate 8.   Oliver Wolcott Gibbs (1822–1908) [(1910), *Nat. Acad. Sci.,*
Biog. Mem. **7**, 1]

first original contribution to this field by discovering a new cobalt–ammonia cation which he obtained by passing oxides of nitrogen into solutions of Genth's compounds. The salts

$$\left[\begin{array}{c} NO_2 \\ H_3N \diagdown \mid \diagup NH_3 \\ Co \\ H_3N \diagup \mid \diagdown NH_3 \\ NH_3 \end{array} \right]^{2+} X_2^-$$

are now known as nitropentaamminecobalt(III) or xanthocobalt compounds. The new cation differed from those previously described in that it contained nitrogen dioxide as well as ammonia and cobalt.

It was not until 1856, however, that Gibbs and Genth's joint results appeared in a 67-page monograph (Gibbs and Genth, 1856a) published by the Smithsonian Institution and reprinted in the *American Journal of Science* (Gibbs and Genth, 1856b, 1857b). After a discussion of their analytical methods, Gibbs and Genth described in detail the preparation, properties, analytical data and reactions of 35 salts of four cobalt–ammine cations: (1) Genth's roseocobalt or aquapentaamminecobalt(III), $[Co(NH_3)_5 H_2 O] X_3$; (2) purpurecobalt or chloropentaamminecobalt(III), $[Co(NH_3)_5 Cl] X_2$:

$$\left[\begin{array}{c} Cl \\ H_3N \diagdown \mid \diagup NH_3 \\ Co \\ H_3N \diagup \mid \diagdown NH_3 \\ NH_3 \end{array} \right]^{2+} X_2^-$$

described by the Frenchman Frédéric Claudet (1851); (3) Genth's luteocobalt or hexaamminecobalt(III), $[Co(NH_3)_6] X_3$, and (4) Gibbs' xanthocobalt or nitropentaamminecobalt(III), $[Co(NH_3)_5 NO_2] X_2$. In eleven cases, they also reported detailed crystallographic data. Here, for the first time, roseo and purpureo compounds were clearly differentiated although Gibbs and Genth erred in supposing them to be isomeric. They pointed the way to the future by correctly predicting coordination compounds in which one or more equivalents of ammonia are replaced by an equal number of equivalents of an organic amine as well as compounds in which cobalt could be replaced by other metals. According to George F. Barker, Gibbs and Genth's elaborate and extended memoir 'has always ranked among the highest chemical investigations ever made in this country'. They concluded their memoir with the statement, 'we invite the attention of chemists to a class of salts which for beauty of form and color, and for abstract theoretical interest, are almost unequalled either among organic or inorganic compounds'.

During the next four decades, their invitation was accepted by a number of chemists too numerous to mention in detail here. Gibbs and Genth thus deserve credit not only for the first distinct recognition of the existence of series of perfectly well-defined and crystallized cobalt—ammine salts but also for attracting chemists to a field which today is experiencing a renaissance of activity. Their experimental results of the early 1850s form a direct transitional link from the primitive, qualitative observations of Libavius, Diesbach and Tassaert to the most recent of sophisticated, quantitative, contemporary investigations. Now that we have concluded our survey of historically important coordination compounds, we shall examine in the next two chapters some of the theories advanced with various degrees of success to explain the constitution and configuration of these colorful and intriguing compounds.

5

Theories of coordination compounds

In most fields of science, theory generally lags behind practice. In other words, sufficient experimental data must be accumulated before attempts are made to explain these experimental facts and to predict new phenomena. As we have seen, during the first half of the nineteenth century, discoveries of coordination compounds were few, sporadic and often accidental, and it was not until after Gibbs and Genth's classic memoir of 1856 that chemists began to devote themselves in earnest to a systematic study of this field. We might therefore think that few theories of coordination compounds were advanced until late in the second half of the nineteenth century, but this was not the case. In coordination chemistry, the lag of theory behind practice was not a great one because of the tremendous importance of coordination compounds to the general question of chemical bonding. In the words of Alfred Werner himself, 'Of all inorganic compounds, [metal–ammines] are best suited to the solution of constitutional problems. . . . it was through the investigation of metal–ammines that the decisive basic principles involved in the constitutional conception of inorganic compounds could first of all be clearly recognized'.

GRAHAM'S AMMONIUM THEORY
(Kauffman, 1972e, 1974e, 1976b)

As the number of known coordination compounds increased, theories to explain their constitution were devised. The Scottish chemist Thomas Graham (1805–1869) (Plate 9) is credited with originating the first theory of metal–ammines, the so-called ammonium theory, in which metal–ammines are considered as substituted ammonium compounds. In 1837, in his book *Elements of Chemistry*, Graham attempted to

Plate 9. Thomas Graham (1805–1869) [(1934), *J. Chem. Educ.* **11**, 281]

explain the consitution of compounds such as diammine copper(II) chloride by the formula

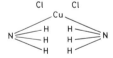

Because of the close analogy between copper and hydrogen, he suggested that two hydrogen atoms, one from each ammonia molecule, had been displaced by the copper atom. He thus viewed the salt as a 'chloride of cuprammonium' similar to the 'chloride of ammonium' (Graham, 1837). He thus accounted for the strong retention of the ammonia in the complex. The theory shows a remarkably close similarity to the modern Lewis acid–base* approach to the formation of coordinate covalent bonds. Despite the fact that Graham's ammonium theory could be applied only when the number of ammonia molecules in the coordination compound was equal to the electrovalence of the metal, it met with a fair degree of success and was generally accepted until Werner's time, largely because of the modifications of it that were proposed by other chemists such as Gerhardt (1850), Wurtz (1850), Reiset (1844), Von Hofmann (1851), Weltzien (1856) and Boedecker (1862).

In 1850 the French chemist Charles Frédéric Gerhardt (1816–1856) applied Graham's idea to the platinum–ammines, which he regarded as ammonia molecules in which different amounts of hydrogen had been replaced by platinum. He considered Reiset's first and second chlorides as compounds of hydrogen chloride with ammonia in which hydrogen had been replaced by *platinosum* (divalent platinum):

Reiset's First Chloride (modern $[Pt(NH_3)_4]Cl_2$): $PtCl + 2NH_3 = N_2H_5Pt + HCl$.

Reiset's Second Chloride (modern *trans*-$[Pt(NH_3)_2Cl_2]$): $PtCl + NH_3 = NH_2Pt + HCl$.

Gerhardt designated NH_2Pt as *platosamine* and N_2H_5Pt as *diplatosamine.*

Another Frenchman, Charles-Adolphe Wurtz (1917–1884), agreed with this formulation but designated NH_2Pt as *platiniaque* and N_2H_5Pt as *platinamine*. Otherwise, Gerhardt's view found few adherents.

In 1844 Jules Reiset (1818–1896), whom we encountered in the last chapter, was the first to suggest that the hydrogen atoms of an ammonium group can be replaced not only by metals but also by other ammonium groups, i.e. hydrogen can be substituted partially by metal and partially by ammonium. According to his formulation of the decomposition of his second chloride (*trans*-$[Pt(NH_3)_2Cl_2]$): $3(NH^3PtCl) = 2(ClH,NH^3) + ClH + NPt^3$, he considered that NPt^3 probably existed, if only ephemerally. He did not agree with Berzelius' view (see pp. 72–74) of platinum–ammines as salts of conjugated ammonia. Instead, he considered it simpler to assume that his chlorides were true compounds, produced by the most intimate union of ammonia with the metal oxide and possessing the properties of all other bases.

In 1851 August Wilhelm von Hofmann (1818–1892), the famous German chemist who spent a number of years in England, specifically applied Reiset's views to the complexes of platinum and cobalt, but in his contribution on the volatile organic bases he frankly confessed that, in his enthusiasm for the ammonium theory, his formulations went somewhat beyond the experimental facts. He represented luteocobalt chloride, $[Co(NH_3)_6]Cl_3$, as

$$Co \left(\begin{array}{c} NH_2{-}NH_4 \\ | \\ Cl \end{array} \right)_3 .$$

and such formulations enjoyed widespread recognition and approval until 1886, when Sophus Mads Jørgensen discovered coordination compounds not explicable according to this approach. These compounds were formed by the salts of silver, copper and platinum with pyridine, a tertiary amine that does not contain any replaceable hydrogen atoms. Such

Plate 10. Jöns Jacob Berzelius (1779–1848) [Sachtleben, R. and Hermann, A. (1960), *Von der Alchemie zur Grossynthese: Grosse Chemiker*, p. 51, Ernst Battenberg Verlag, Munich]

formulations also did not explain why removal of one of the ammonia molecules completely changed the function of one of the chlorine atoms, rendering it nonionizable, a phenomenon later explained elegantly by Werner's coordination theory.

BERZELIUS' CONJUGATE THEORY
(Kauffman, 1974e, 1976b)

Shortly after Graham proposed his ammonium theory (1837), the great Swedish chemist Jöns Jacob Berzelius (1779–1848) (Plate 10) came forth with his own attempt to explain the constitution of coordination compounds – his so-called conjugate theory. Berzelius, the supreme authority and arbiter of matters chemical during the first half of the nineteenth century, dominated this period with his dualistic electrochemical theory, which he first enunciated in 1811. According to this theory, elements are either electropositive or electronegative, and chemical combination results from the mutual neutralization of opposite charges. However, the compounds thus formed are not necessarily neutral because the opposite charges of the combining atoms are not necessarily equal in magnitude. For example, copper and oxygen combine to form copper oxide, and sulfur and oxygen form sulfur trioxide:

$$\overset{+}{Cu} + \overset{-}{O} \rightarrow CuO \text{ (slightly +)} \qquad \text{primary compound}$$

$$\overset{+}{S} + \overset{-}{3O} \rightarrow SO_3 \text{ (slightly -)} \qquad \text{primary compound}$$

Furthermore, since copper oxide and sulfur trioxide are not neutral, they can combine with each other to form copper sulfate:

$$\overset{+}{CuO} + \overset{-}{SO_3} \rightarrow CuO \cdot SO_3, \text{ i.e. } CuSO_4 \text{ (slightly +) secondary compound}$$

Even this salt was not necessarily neutral, and it could combine further to form hydrates, which Berzelius classified as 'compounds of higher order', a term later used by Blomstrand and Werner to describe coordination compounds:

$$\overset{+}{CuO \cdot SO_3} + 5H_2O \rightarrow CuO \cdot SO_3 \cdot 5H_2O, \text{ i.e. } CuSO_4 \cdot 5H_2O$$
$$\text{tertiary compound}$$

In 1841 Berzelius proposed his conjugate theory, using terms and ideas (*corps copulés*) that he borrowed from the French chemist Charles Gerhardt (Berzelius, 1841). According to this theory, he viewed metal–ammines as conjugated or copulated compounds consisting of ammonia and a conjugate* or copula*. The conjugate cannot be removed by reaction with an acid and neither increases nor decreases the saturation capacity of a base. In other words, a metal in conjugation with ammonia is still capable of combining with other substances. For example, the formula which Berzelius assigned to Reiset's First Chloride ($[Pt(NH_3)_4]Cl_2$) was:

$$Pt\overline{NH}^2 \cdot NH^4 \cdot Cl$$

i.e. $Pt(NH_2 \cdot NH_4 \cdot Cl)_2$ or $PtN_2H_4 \cdot N_2H_8Cl_2$. He considered it to be a conjugated compound of *platinamide*, $Pt\overline{NH_2}$, with the conjugate $N_2H_8Cl_2$ consisting of two molecules of ammonium chloride. Berzelius used barred atoms to represent two atoms or doubled atoms.

Berzelius had introduced his theory of copulae in order to reconcile his electrochemical theory with the new phenomena obtained in the organic field which seemed incompatible with it. In short, the great liberal of the early 1800s had become a conservative and even reactionary force during his later years. As time passed, the number of supporters of his theory diminished, and the number of its opponents increased. More and more complicated and improbable formulae became necessary to make the theory agree with experimental facts. For example, the French chemist

Auguste Laurent (1808–1853) ridiculed Berzelius' device as follows: 'What then is a copula? A copula is an imaginary body, the presence of which disguises all the chemical properties of the compounds with which it is united'. In short, Berzelius' stratagem of regarding some compounds as conjugated compounds bore little resemblance to reality in many cases, and although Claus in 1854 and Blomstrand in 1869 in his chain formulae attempted to revive and modify Berzelius' ideas, Berzelius' theory otherwise was of little value.

CLAUS' AMMONIA THEORY (Kauffman, 1976b)

The next major theory of metal–ammines that we shall examine was proposed by Carl Ernst Claus (1796–1864), also known by the Russian name of Karl Karlovich Klaus, Professor at the University of Kazan and later Professor of Chemistry at the University of Dorpat (Plate 11) (Kauffman, 1971, 1974e). In 1854 Claus rejected the ammonium theory and suggested a return to Berzelius' view of complexes as conjugated compounds. According to Claus, the platinum–ammines should be compared not with ammonium salts or with ammonium hydroxide as advocated by adherents of the ammonium theory but with metal oxides. He designated the coordinated ammonia molecule as 'passive, in contrast to

Plate 11. Karl Karlovich Klaus (Carl Ernst Claus) (1796–1864)
[Courtesy, Dr L. B. Hunt, Johnson, Matthey & Co. Ltd, London]

the active, alkaline state in the ammonium salts, where it can easily be detected and replaced by other bases'.

Claus' propositions were summarized as three statements in his more widely read paper of 1856, 'Ueber die Ammoniummoleküle der Metalle' (Claus, 1856):

(1) If several equivalents of ammonia (from two to six) combine with an equivalent of certain metal chlorides, neutral substances are formed, in which the basic property of ammonia has been destroyed and simultaneously the ammonia can be neither detected by the usual methods nor eliminated by double decomposition.

(2) If the chlorine in these compounds is replaced by oxygen, strong bases are obtained, whose saturation capacity is always determined by the oxygen equivalents contained in them but not by the number of equivalents of ammonia present in them.

(3) The number of equivalents of ammonia entering into these substances is not a random one; as is evident from a number of facts, it is determined by the number of equivalents of water contained in the hydrates of the metal oxides which can enter into such compounds along with the ammonia.

Claus' first postulate was vigorously attacked in the same year by Carl Weltzien (1813–1870), famed as the organizer of the Karlsruhe Congress, who considered the term 'passive' to be indefinite and confusing. Weltzien, one of the leading proponents of the rival ammonium theory, insisted that every part of a molecule affects every other part and therefore no part can be considered passive.

In 1862 Hugo Schiff (1834–1915), Professor of Chemistry at the University of Florence and discoverer of the so-called Schiff bases, attacked not only Claus' first postulate but also his second. Schiff pointed out that the oxides of the metal–ammines were much stronger bases than the metal oxides themselves. This criticism seems to imply that Schiff was confusing the strength of a base, i.e. the hydroxide ion concentration of its solution, with its saturation capacity, i.e. the number of equivalents of acid with which it could combine – a distinction that is made today even in elementary chemistry courses. However, we should remember that our hindsight is always far better than our foresight. If we view the works of scientists of the past in the light of modern knowledge, we are unfairly belittling their achievements. Therefore, in evaluating events in the history of chemistry, as we are doing in this book, we must try to preserve a sense of historical perspective.

Claus' third postulate, that is, the close parallel between metal salt hydrates and metal–ammines, was attacked on the grounds that many hydrates were known for which corresponding ammines were unknown. All of Claus' three postulates reappeared in modified form almost four decades later in Alfred Werner's coordination theory. In fact, Claus' third postulate closely adumbrates Werner's concepts of the *Koordinationszahl* (coordination number) and of the *Übergangsreihe* (transition series) between metal–ammines and metal salt hydrates.

CONSTANT VALENCY AND KEKULÉ'S 'MOLECULAR COMPOUNDS' (Kauffman, 1972c)

The next theory of coordination compounds that we shall examine was also applicable to a wide variety of substances. It was proposed by none other than the patriarch of structural organic chemistry, August Kekulé, Professor of Chemistry at the Universities of Ghent and Bonn (1829–1896) (Plate 12).

At about the time that Claus proposed his ammonia theory of metal–ammines, the concept of valence* was being formulated and developed by a number of chemists – in particular, Kekulé, Frankland, Williamson, Odling, Kolbe and Couper. During the late nineteenth and early twentieth centuries the principal difficulty in the field of valence was its applica-

Plate 12. August Kekulé (1829–1896) [Sachtleben, R. and Hermann, A. (1960), *Von der Alchemie zur Grossynthese: Grosse Chemiker*, p. 36, Ernst Battenberg Verlag, Munich]

tion to *all* types of chemical compounds, and one of the main contro-
versies involved whether or not a given element could possess more than
one valence. Inasmuch as coordination compounds pose a number of
basic constitutional problems, it is not surprising that they became involved in the question of variable vs constant valence.

What may come as a surprise to some readers, however, is that Kekulé's
valence theory, which was so flexible and fruitful in the realm of organic
chemistry, proved to be a virtual straitjacket when applied to inorganic
compounds. Yet, by his own admission, Kekulé's concept of constant
valence proved, in his own words, 'embarrassing to the chemist'. However,
instead of abandoning this obviously untenable belief, he compounded
his error by invoking a still more unsatisfactory concept in order to maintain it, namely, the concept of 'molecular compounds' (Kekulé, 1864).

Most of the pioneers in the theory of valence, such as the Englishman
Edward Frankland (1825–1899) and the Scot Archibald Scott Couper
(1831–1892), readily admitted the possibility of variable valence. In
other words, they felt that a given element could exhibit one valence in
one compound and a different valence in another compound. On the
other hand, Kekulé, from his first statements on the self-linking of carbon
atoms in 1858 until his death in 1896, adopted and rigidly adhered to the
principle of constant valence. In spite of the mass of data that soon
accumulated to contradict such a simple and admittedly attractive
assumption, Kekulé dogmatically insisted that atomicity, which was the
term that he used for valence, was, in his own words, 'a fundamental
property of the atom which is just as constant and unchangeable as the
atomic weight itself'. The simplicity of this principle, however, was more
than outweighed by the complicated and unrealistic formulae required
in order to maintain it, and eventually the stubborn Kekulé stood virtually
alone in its defense. Once again, as we have seen with Berzelius, the liberal
of one generation had become the conservative of the next.

Kekulé's dichotomy of compounds into 'atomic compounds' and
'molecular compounds' was an attempt to buttress his theory of constant
valence. According to Kekulé:

Compounds in which all the elements are held together by the
affinities of the atoms which mutually saturate one another
could be called *atomic compounds.* They are the only ones
which can exist in the vapor state . . . . We must distinguish a
second category of compounds that I shall designate *molecular
compounds* (Kekulé, 1864).

A few examples should suffice to illustrate Kekulé's concept of 'molecular compounds'. Since Kekulé regarded the valences of nitrogen,

phosphorus and cobalt as invariably three, and of copper as invariably two, he was forced to consider phosphorus(V) chloride, ammonium chloride, copper(II) sulfate pentahydrate and hexaamminecobalt(III) chloride as 'molecular compounds' with the formulae shown on the left hand side of the following equations:

PCl$_5$
(phosphorus(V) chloride)

$$PCl_3 \cdot Cl_2 \xrightarrow{\Delta} PCl_3 + Cl_2$$

NH$_4$Cl
(ammonium chloride)

$$NH_3 \cdot HCl \xrightarrow{\Delta} NH_3 + HCl$$

CuSO$_4 \cdot 5H_2O$
(copper(II) sulfate pentahydrate)

$$CuSO_4 \cdot 5H_2O \xrightarrow{\Delta} CuSO_4 + 5H_2O$$

[Co(NH$_3$)$_6$]Cl$_3$
(hexaamminecobalt(III)
chloride)

$$CoCl_3 \cdot 6NH_3 \xrightarrow{\Delta} \text{no reaction}$$

NaOH   H$_2$SO$_4$

[Co(NH$_3$)$_6$](OH)$_3$ [Co(NH$_3$)$_6$]$_2$(SO$_4$)$_3$
(no Co(OH)$_3$ formed)   (no NH$_4^+$ salt formed)

Today, Kekulé's mysterious noncommital dot has all but disappeared in writing the formulae of coordination compounds. When we occasionally still use it to write the formulae of metal salt hydrates or of hydrochlorides of organic bases, we unwittingly invoke the ghost of Kekulé and his now defunct doctrine of constant valency.

A page from a holograph book of Alfred Werner's elementary chemistry notes, in which we see PCl$_5$ formulated as a 'molecular compound' in accordance with Kekulé's doctrine of constant valency, appears in Plate 13. This 127-page book in Werner's handwriting dates from 1883–84 when he was between 17 and 18 years old. A decade later, in his coordination theory of 1893, Werner was destined to offer an alternative and much more satisfactory explanation for the constitution and configuration of what were then called 'molecular compounds'.

In a sense, Kekulé's concept of 'molecular compounds' was a revival of Berzelius' dualistic theory whereby 'secondary compounds' (in Kekulé's terminology, 'atomic compounds') containing a small excess of electrical charge could still combine with other 'secondary compounds' containing a small excess of opposite charge to form 'tertiary compounds' (in Kekulé's terminology, 'molecular compounds'). At most, Kekulé's artificial division of compounds into 'atomic compounds', which obeyed the rules of classical valence theory, and into 'molecular compounds', which did not obey these rules, had some limited value as a formal classification. However, in no way did it explain the nature or operation of the forces

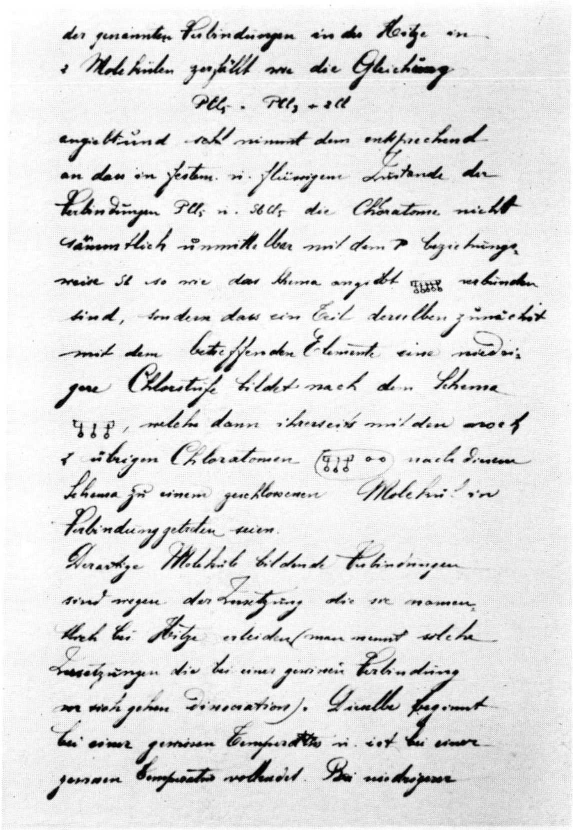

Plate 13. An early notebook of Alfred Werner's [*Einleitung in die Chemie, Mulhouse 1883–84*, p. 111]

involved in the formation of 'molecular compounds' except to assume that the forces were acting between molecules rather than between atoms.

Since the forces acting between molecules were supposedly weaker than the forces acting between atoms, according to Kekulé, the resulting 'molecular compounds' should be less stable than 'atomic compounds'. Indeed, some of the substances of limited thermal stability cited by Kekulé as prototypes of 'molecular compounds', such as phosphorus(V) chloride, ammonium chloride and copper(II) sulfate pentahydrate, did decompose in the vapor state (see equations on p. 78). However, this was a relative rather than an absolute phenomenon. For example, under certain conditions, such as the use of lower temperatures or the addition of decomposition products, many of these substances can be vaporized without decomposition. Therefore, many chemists began to regard Kekulé's classification as meaningless. The great Russian chemist, Dmitriǐ Ivanovich Mendeleev, the discoverer of the periodic law, wrote:

> Kekulé's division of chemical compounds into 'atomic' and 'molecular' types is artificial, arbitrary and unsound . . . . No practical test exists by which the two categories may be sharply separated.

But Kekulé's stability criterion, or to be more accurate, instability criterion, failed completely in the case of many coordination compounds, especially the metal–ammines, which were classified as 'molecular compounds' by sheer dint of necessity even though they were extremely resistant to heat and chemical reagents. For example, although hexa-amminecobalt(III) chloride contains ammonia, it neither evolves this ammonia on mild heating nor does it react with acids to form ammonium salts. Also, despite its cobalt content, addition of a base to its aqueous solution fails to precipitate hydrated cobalt(III) hydroxide (see equations on p. 78). It remained for Alfred Werner to explain successfully the constitution of such compounds.

By the time that Werner entered the scene to deliver the final *coup de grâce* to Kekulé's idea of constant valence (1893), amusing but admittedly logical definitions such as the following were widespread: 'Atomistic compounds are those which can be explained by constant atomicity. All others are to be conceived as molecular compounds. . . .'

> Whatever in chemistry could not be defined was regarded as a molecular compound, a concept in which all kinds of things could be included without thereby achieving the slightest clarification. With molecular compounds, conditions were about the same as with animal instinct.

Werner agreed with this evaluation by Fritz Reitzenstein (1898) by describing the viewing of certain compounds as molecular compounds as 'substituting a beautiful word for a confused concept' (Werner, 1894, p. 270). On the very first page of his paper on the coordination theory, Werner resolutely cut the Gordian knot that had caused decades of confusion, thus terminating a situation that had clearly become intolerable:

> In view of the present stage of science, it seems inadmissible to classify metal–ammonia salts according to their stability into different classes of compounds, of which the stable ones would

receive atomistic constitutional formulae and the unstable ones so-called molecular formulae; we must look for another principle of subdivision (Werner, 1893, p. 267; Kauffman, 1968).

The coordination theory of Alfred Werner would make the artifices of constant valence and molecular compounds unnecessary and would begin a new era in the chemistry of coordination compounds. However, before considering Werner's coordination theory, we must examine one more theory of coordination compounds.

THE BLOMSTRAND–JØRGENSEN CHAIN THEORY
(Kauffman, 1959, 1960, 1970b, 1975b, 1976b, 1977d)

Whereas Kekulé disposed of complex compounds by banishing them to the limbo of 'molecular compounds', other chemists of the time developed highly elaborate theories in order to explain the constitution and properties of these colorful and intriguing substances. The most successful and widely accepted of such pre-Werner theories was undoubtedly the so-called chain theory, advanced by Christian Wilhelm Blomstrand (1826–1897), Professor of Chemistry and Mineralogy at the University of Lund (Plate 14) (Kauffman, 1975b).

Plate 14. Christian Wilhelm Blomstrand (1826–1897) [Courtesy, Universitetsbiblioteket, Lund, Sweden]

Living in Sweden during a transition period between the older and newer chemistry and being a scientific as well as a political conservative, Blomstrand tried to reconcile Berzelius' old dualistic theory with the newer unitary and type theories. In fact, his best known work, the book *Die Chemie der Jetztzeit* (The Chemistry of Today) in which he first proposed his chain theory, bears the subtitle 'from the viewpoint of the electrochemical interpretation developed from Berzelius' theory' (Blomstrand, 1869). He considered the newer atomic theory to be 'only a consequential development of Berzelius' atomic theory necessarily evoked by the force of many newly discovered facts'.

Despite his conservatism, Blomstrand was opposed to Kekulé's dogma of constant valency, and he tried to establish a sound and complete theory of variable valency. For him, Kekulé's dichotomy of compounds into 'atomic compounds' and 'molecular compounds' was completely unacceptable. In his book of 1869, Blomstrand asserted that 'It has become the principal task of the newer chemistry to explain atomistically, i.e. from the saturation capacity [valence] of the elements, compounds which previously have been conceived of more or less definitely as molecular'. This statement was later chosen by no less an authority on coordination compounds than Alfred Werner as the motto for the frontispiece of his monumental but modestly titled textbook *Neuere Anschauungen auf dem Gebiete der anorganischen Chemie* (Newer Views in the Field of Inorganic Chemistry) (Werner, 1905).

Sophus Mads Jørgensen (1837–1914), Professor of Chemistry at the University of Copenhagen (Plate 15), was Werner's primary scientific adversary (Kauffman, 1959, 1960, 1973i, 1976d). Except for some early isolated research, Jørgensen devoted himself exclusively to investigating the coordination compounds of cobalt, chromium, rhodium and platinum. This work, upon which his fame securely rests, forms an interconnected and continuous chain from 1878 to 1906. Although Werner's ideas eventually triumphed, this in no way invalidated Jørgensen's experimental observations. On the contrary, his experiments, performed with meticulous care, have proven completely reliable. They provided the experimental foundation not only for the Blomstrand–Jørgensen chain theory but for Werner's coordination theory as well.

As a research worker, Jørgensen was methodical, deliberate, painstaking and solitary. Although he could have delegated much routine work to assistants, he insisted on personally performing all his analyses. In fact, he reserved one day a week for this task. Werner, on the other hand, allowed many of the details of his syntheses to be worked out by his assistants or students. Consequently, almost all of Jørgensen's experimental work is reproducible, whereas some of Werner's work in this area leaves much to be desired. In view of Jørgensen's passion for perfection, his research out-

Plate 15. Sophus Mads Jørgensen (1837–1914) [(1954), *Proceedings of the Symposium on Co-ordination Chemistry, Copenhagen, August 9–13, 1953*, p. 14, Danish Chemical Society, Copenhagen]

put was tremendous, and we are indebted to him for many of the basic experimental facts of coordination chemistry.

Now the latter half of the nineteenth century was a period of tremendous progress in organic chemistry, and organic chemistry exerted a predominant influence on other branches of chemistry. Thus Blomstrand suggested that ammonia molecules could link together as $-NH_3-$ chains, analogous to $-CH_2-$ chains in hydrocarbons. These chains involved 'quinquevalent' nitrogen, an idea that Kekulé found anathema. Today, even beginning chemistry students familiar with modern orbital theory are aware that nitrogen, like the other elements of the second period of the periodic table, is capable of forming at most four bonds, but in Blomstrand's time formulae depicting nitrogen atoms with five bonds were quite common.

The number of ammonia molecules associated with the metal, that is, the length of the chain, depended upon the metal and its valence. This point was later accounted for more adequately by Werner's concept of the coordination number. Jørgensen made provision for different reactivities of various atoms and groups. For example, halogen atoms that could not be precipitated immediately by silver nitrate were called 'nearer' and were considered to be bonded directly to the metal atom. Halogen atoms that could be precipitated immediately by silver nitrate were called 'farther' and were considered to be bonded through the ammonia chains.

These two different kinds of bonding were later explained more satis-
factorily by Werner's term 'nonionogenic' and 'ionogenic', respectively,
and by his concepts of inner and outer spheres of coordination. Despite
the chain theory's admitted limitations, it permitted the correlation of a
considerable amount of empirical data.

In 1883, as we have seen, Blomstrand's friend Jørgensen demonstrated
that tertiary amines, which contain no replaceable hydrogen, are capable
of forming compounds that are completely analogous to the metal–
ammines both in their composition and in their properties. Graham's
ammonium theory (pp. 69–72) conceived of metal–ammines as salts in
which some of the hydrogen atoms of the ammonium group were replaced
by metal atoms. Since tertiary amines contain no replaceable hydrogen,
Jørgensen's discovery effectively eliminated the ammonium theory from
serious consideration as an explanation for the constitution of the metal–
ammines. Therefore, chemists at that time were forced to assume the
existence of ammonia chains copied from hydrocarbons or to conceive of
metal–ammines as 'molecular compounds'. Since Kekulé's theory really
explained nothing and only 'substituted a beautiful word for a confused
concept', to quote Werner again, the Blomstrand–Jørgensen chain theory
became the most popular and satisfactory way to accounting for metal–
ammines. It held sway for almost a quarter of a century until it was displaced by Werner's coordination theory.

Wilhelm Ostwald has divided scientific geniuses into two types – the
classic and the romantic. Jørgensen seems the embodiment of the classic
type – the conservative, slow and deep-digging completer who produces
only after long deliberation and who methodically develops a traditional
theory to new consequences. Jørgensen's strong and conservative sense of
history caused him to regard Werner's new theory as an unwarranted
break in the development of theories of chemical structure. He regarded
it as an *ad hoc* explanation insufficiently supported by experimental
evidence.

Although Jørgensen created no new structural theory of his own, he
logically and consistently extended and modified Blomstrand's chain
theory in order to interpret the many new series of complexes that he,
Jørgensen, had succeeded in preparing for the first time. Just as the
medieval astronomers tried to force an explanation for the motion of the
planets in terms of the old geocentric Ptolemaic theory by postulating
more and more complicated epicycles, so did Jørgensen strain to the
breaking point the theory of his mentor Blomstrand in his attempt to
account for his newly prepared coordination compounds from a unified,
theoretical point of view. Finally, in 1893, the Copernican figure of Alfred
Werner appeared on the scene to challenge the old system with a revolutionary new theory based, according to Werner's own admission, upon

the sturdy foundation of Jørgensen's painstaking experimental investigations. Ironic as it may seem, Jørgensen's work bore the seeds of the
Blomstrand–Jørgensen theory's destruction, for, as we shall soon see,
some of the compounds first prepared by Jørgensen later proved instru-
mental in demonstrating the validity of Werner's views. It is tempting
to compare this situation with Priestley's discovery of oxygen, which led
to Lavoisier's classic experiments on the nature of combustion and to the
subsequent collapse of the phlogiston theory. However, unlike Priestley,
who staunchly defended the phlogiston theory until his death, Jørgensen
finally became convinced of the correctness of Werner's theory and
acknowledged its worth. Since in the next chapter we shall compare in
some detail the predictions of the Blomstrand–Jørgensen chain theory
with those of the Werner coordination theory, we shall postpone a
detailed consideration of the chain theory until we have examined the
basic postulates of the coordination theory.

6

Alfred Werner's coordination theory and the Werner–Jørgensen controversy

In 1893 a comparatively unknown 26-year-old *Privat-Dozent* at the Eidgenössisches Polytechnikum in Zürich came forth to challenge and discard the confining rigidities of both the Kekulé constant valence theory and the Blomstrand–Jørgensen chain theory. Like a modern Alexander the Great, Alfred Werner cut the Gordian knot that for decades had caused confusion in structural inorganic chemistry. The era of coordination chemistry had begun.

The circumstances surrounding the creation of Werner's coordination theory provide us with a classic example of the 'flash of genius' that ranks with August Kekulé's famous dreams of the benzene ring and of the self-linking of carbon atoms. At the time, Werner's primary interest lay in the field of organic chemistry, and his knowledge of inorganic chemistry was extremely limited. Perhaps there is some truth after all in Albert Einstein's statement that 'imagination is more important than knowledge', for one night in late 1892 Werner awoke at 2 a.m. with the solution to the problem of the constitution of 'molecular compounds', which had come to him like a flash of lightning. He arose from his bed and wrote furiously and without interruption. By 5 p.m. of the following day he had finished his most famous paper entitled 'Beitrag zur Konstitution anorganischer Verbindungen' (Contribution to the Constitution of Inorganic Compounds) (Werner, 1893; Kauffman, 1966b, c, 1976b, e).

This event might lead us to consider Werner to be the prototype of Ostwald's second type of genius – the romantic – the liberal, even radical, impulsive and brilliant initiator who produces prolifically and easily during his youth, in short, the exact opposite of his adversary Jørgensen. Yet Werner's personality was too complex and self-contradictory to be accommodated by Ostwald's oversimplified dichotomy. At the time of its inception, Werner's theory was largely without experimental verification. The data cited by Werner in support of his ideas had been obtained by the painstaking efforts of others, especially of Jørgensen. After giving birth to the coordination theory, the typical romantic genius of the stereotype would have diverted his attention elsewhere, possibly to devising new, additional theories, and left to others the long and arduous task of accumulating the experimental data necessary for its rigorous proof. But Werner combined the impulsive, intuitive and theoretical brilliance of the romantic with the thorough, practical and experimental persistence of the classicist. Firmly convinced of the correctness of his views, he devoted the remainder of his career to an almost unprecedented series of experimental researches which explored nearly every conceivable aspect of coordination chemistry and simultaneously verified his original theory in virtually every particular.

THE COORDINATION THEORY

We are now ready to examine the basic postulates of the coordination theory. To some extent, this will constitute a repetition of some of the definitions of terms and concepts mentioned in Chapter 1. However, our outlook here will differ from our earlier viewpoint. Instead of regarding Werner's concepts as proven facts, let us try to go back in time and compare as objectively as possible the Blomstrand—Jørgensen chain theory with Werner's coordination theory. If we can regard ourselves as contemporary spectators, we may be able to see the advantages and disadvantages of each theory instead of regarding the view currently accepted, namely, Werner's view, as correct and any other views as hopelessly naive. If history teaches us anything, it teaches us that the latest view is not always the best and that change is not always progress.

   In his revolutionary theory, which marked an abrupt break with the classical theories of valence and structure, Werner postulated two types of valence — *Hauptvalenz*, primary or ionizable valence, and *Nebenvalenz*, secondary or nonionizable valence. According to Werner, every metal in a particular oxidation state, that is, with a particular primary valence, also has a definite *coordination number*, that is, a fixed number of secondary valences that must be satisfied. Now, whereas primary or ionizable valences can be satisfied only by anions, secondary or nonionizable valences can be satisfied not only by anions but also by neutral molecules containing donor atoms such as nitrogen, oxygen, sulfur and phosphorus. Typical ligands which can be coordinated to the central atom in this manner include ammonia, organic amines, water, organic sulfides and phosphines. The secondary valences are directed in space around the central metal atom, and the combined aggregate forms a 'complex', which usually exists as a discrete unit in solution. Typical configurations are octahedral for coordination number six and square planar or tetrahedral for coordination number four.

In order to appreciate adequately the magnitude of both Jørgensen's and Werner's achievements, we must take into account the confused state of affairs in which chemistry floundered during most of the nineteenth century. Various rival systems of chemistry flourished. The dualistic theory of Berzelius, which hitherto had been quite successful in the formulation of inorganic compounds, was falling into disrepute as a result of the inroads of the new organic chemistry. No clear distinction was made between equivalent, atomic and molecular weights. It was only in 1858 that Cannizzaro's revival of Avogadro's hypothesis marked the beginnings of a consistent atomic weight scale. When Cannizzaro spoke at Karlsruhe, Svante Arrhenius had not yet been born. Indeed, years of proselytizing by Arrhenius, Van't Hoff and Ostwald were to be necessary before the electrolytic dissociation theory was finally accepted by the scientific world. Thus Werner's view of the two types of linkage, ionizable ('ionogenic') and nonionizable ('nonionogenic'), did much to clarify ideas of chemical bonding a generation before the views of Kossel and Lewis in 1916 led to our present concepts of ionic and covalent bonding.

As mentioned in the last chapters, today any chemist familiar with modern orbital theory knows that nitrogen can form at most only four bonds. Armed with such knowledge, he might scoff at the apparent naïveté of Blomstrand and Jørgensen whose structural formulas involved chains of ammonia molecules containing quinquevalent nitrogen. To view the works of great men of the past in the light of modern knowledge is perhaps to belittle their achievements, and in evaluating and comparing the work of Jørgensen and Werner we must guard against this general tendency.

In the following comparison between the chain theory and the coordination theory, we shall confine ourselves to the most common type of complexes, namely the octahedral hexacovalent ammines of cobalt(III). Although we are concentrating on coordination number six, please bear in mind that Werner used similar arguments to prove the constitution and configuration for compounds of coordination number four. These will be considered in Chapter 7 (pp. 162–167). Our survey here, which includes only the most important aspects of the controversy, will be organized on the basis of compound type, that is, in a logical rather than strictly chronological sequence. First we will consider type MA_6 in which the coordination number of the central metal atom is satisfied by six ammonia molecules. We shall then proceed to replace these ammonia molecules one at a time with other groups (Kauffman, 1959, 1960, 1966b, 1967c, 1976b, 1977c).

The acknowledged test of a scientific theory is its ability to explain known facts and to predict new ones. In examining the comparative successes of the chain theory and the coordination theory in meeting these criteria, we shall examine the metal–ammines under two aspects: (1) *constitution*, that is, the manner of bonding of the constituent atoms and groups, and (2) *configuration*, that is, the spatial arrangement of these atoms and groups.

CONSTITUTION OF COBALT–AMMINES

Type MA_6 —Hexaammines (luteo salts), $[M(NH_3)_6]X_3$
Luteo salts may be regarded as the parent compounds of all other hexa-covalent complexes, which can be derived from them by replacing the ammonia molecules with other groups.

During the first half of the nineteenth century, measurement of vapor density was the only method for determination of molecular weights. Until the classical studies of François Marie Raoult (1830–1901) and Jacobus Henricus van't Hoff (1852–1911) in about 1882 on colligative properties of solutions, no reliable method existed for the determination of molecular weights of nonvolatile compounds. By analogy with volatile Fe_2Cl_6, nonvolatile cobalt(III) chloride was thought to have the composition Co_2Cl_6, and hence cobalt–ammines were considered dimers. Jørgensen (1890a) and Petersen (1892) deduced evidence for monomeric molecular weights by freezing-point and conductivity measurements of such solutions, and Blomstrand's original formulas were halved. Thus, luteo cobaltic chloride, originally written $Co_2Cl_6 \cdot 12NH_3$, was henceforth written $CoCl_3 \cdot 6NH_3$. The concept of octahedral configuration based on coordination number six was a fundamental postulate of Werner's theory from its inception. It is possible that without Jørgensen's halving of Blomstrand's formulae, this theory might never have been conceived. The close relationship and interdependence of the various branches of chemistry is thus underscored.

Luteo salts can be prepared by heating purpureo (pentaammine) salts with concentrated ammonia; for example:

$$[Co(NH_3)_5Cl]Cl_2 + NH_3 \xrightarrow{\Delta} [Co(NH_3)_6]Cl_3\dagger$$

Luteo cobaltic chloride was found to be a stable yellow-orange compound (Jørgensen, 1899a). In solution, all the chlorine is immediately precipitat-

† Equations are given in the modern form, i.e. in accordance with Werner's coordination theory.

ed by silver nitrate. Although ammonia is a base, treatment of luteo cobaltic chloride with hydrochloric acid at 100 °C does not remove any ammonia. Furthermore, treatment of the solid with sulfuric acid does not remove any ammonia, but yields the compound $Co_2(SO_4)_3 \cdot 12NH_3$, i.e. the chlorine atoms are replaced by sulfate groups:

$$2[Co(NH_3)_6]Cl_3 + 3H_2SO_4 \longrightarrow [Co(NH_3)_6]_2(SO_4)_3 + 6HCl \uparrow$$

Clearly, some sort of very stable metal–ammonia bonding but much less stable metal–chlorine bonding is indicated in luteo cobaltic chloride. Blomstrand proposed the symmetrical formula

$$Co_2 \begin{cases} NH_3-NH_3-Cl \\ NH_3-NH_3-Cl \\ NH_3-NH_3-Cl \\ NH_3-NH_3-Cl \\ NH_3-NH_3-Cl \\ NH_3-NH_3-Cl \end{cases}$$

On heating, however, one-sixth of the ammonia is lost, and only two-thirds of the chlorine in the resulting purpureo cobaltic chloride can now be precipitated by silver nitrate:

$$[Co(NH_3)_6]Cl_3 \xrightarrow{\Delta} [Co(NH_3)_5Cl]Cl_2 + NH_3 \uparrow$$

Removal of two ammonia molecules from Blomstrand's formula for the luteo salt results in a structure which does not sufficiently account for the great difference between the two types of chlorine atoms in the purpureo salt. Therefore, Jørgensen proposed the following symmetrical formula for the luteo salt

$$Co_2 \begin{cases} NH_3-Cl \\ NH_3-NH_3-Cl \\ NH_3-NH_3-NH_3-Cl \\ NH_3-NH_3-NH_3-Cl \\ NH_3-NH_3-Cl \\ NH_3-Cl \end{cases}$$

postulating that halogen atoms that are bonded to the metal atom through other groups such as ammonia can be precipitated by silver nitrate, while those bonded directly to the metal atom cannot (Jørgensen, 1887). Since he later regarded four as the maximum number of ammonia molecules that could enter into a chain and since he regarded such a chain as a particularly stable arrangement, Jørgensen (1894a) subsequently modified this formula to

$$Co_2 \begin{cases} NH_3-Cl \\ NH_3-Cl \\ NH_3-NH_3-NH_3-NH_3-Cl \\ NH_3-NH_3-NH_3-NH_3-Cl \\ NH_3-Cl \\ NH_3-Cl \end{cases}$$

Henceforth, all our representations of Jørgensen's structural formulae will show this four-ammonia chain although his original proposals were slightly different. Both Jørgensen's and Werner's structures

$$Co \overset{\displaystyle \nearrow NH_3-Cl}{\underset{\displaystyle \searrow NH_3-NH_3-NH_3-NH_3-Cl}{-NH_3-Cl}} \qquad [Co(NH_3)_6]Cl_3$$

Jørgensen   Werner

are compatible with the experimental observations. (For simplicity, monomeric formulae will be used for the remainder of this discussion although they were not used by Jørgensen until 1890.) Their formulae, however, differ in that Jørgensen regarded the chlorine atoms as attached to the metal atom through ammonia molecules, while Werner regarded them as ionogenic and outside the coordination sphere. The ionic nature of salts in the solid state, now confirmed experimentally by X-ray diffraction and other methods, was then unknown. Werner also regarded the central metal atom and the ammonia molecules as comprising a discrete unit, a complex cation. Such a structure should yield four ions in solution, and this was later confirmed by conductivity studies (Werner and Miolati, 1893, 1894).

A complete series of yellow hexaamminechromium(III) salts, $[Cr(NH_3)_6]X_3$, strictly analogous to those of cobalt(III), as well as other chromium(III)–ammines were discovered by Jørgensen. He also prepared and investigated many hexaammines of rhodium(III), which were found to have properties similar to those of cobalt(III) and chromium(III).

Type MA_5B – pentaammines, $[M(NH_3)_5 X^{-n}]Y_{3-n}$
These compounds may be regarded as luteo salts in which one-sixth of the ammonia has been replaced by another group. Depending upon the replacing group, this type can be subdivided into several series.

(a) Purpureo salts. These are pentaammines in which the replacing group is sulfate, nitrate, oxalate or a halogen. The term purpureo (purple) is derived from the color of purpureo cobaltic chloride, $[Co(NH_3)_5 Cl]Cl_2$, the substance with which Jørgensen began his research on complexes in

1878. This compound is formed by heating luteo cobaltic chloride (Jørgensen, 1899):

$$[Co(NH_3)_6]Cl_3 \xrightarrow[>100\ °C]{\Delta} [Co(NH_3)_5Cl]Cl_2 + NH_3 \uparrow$$

and thus purpureo salts were regarded by Jørgensen as luteo salts in which one-sixth of the ammonia had been replaced by halogen. The original luteo salt may be regenerated by treating the purpureo salt with concentrated ammonia. The ammonia molecules are strongly bonded to the cobalt atom in the purpureo salt as shown by the fact that ammonia is not evolved even on heating to 100 °C. Furthermore, Jørgensen showed that cold concentrated sulfuric acid does not react with the ammonia in the salt, but yields the compound $[Co(NH_3)_5Cl]SO_4$, which, although it contains chlorine, gives no immediate precipitate with silver nitrate:

$$[Co(NH_3)_5Cl]Cl_2 + H_2SO_4 \rightarrow [Co(NH_3)_5Cl]SO_4 + 2HCl$$

He also found that only two-thirds of the chlorine in the original salt can be immediately precipitated by silver nitrate, while the remaining third is precipitated only on long boiling (Jørgensen, 1878, 1899).

To account for this difference in reactivity, Jørgensen suggested, as did Werner after him, that the 'unreactive' or 'masked'* chlorine was bound directly to the metal atom. The structures proposed by these two investigators,

Co—NH$_3$—Cl structure with Cl above and NH$_3$—NH$_3$—NH$_3$—NH$_3$—Cl below, labeled Jørgensen; and $[Co(NH_3)_5Cl]Cl_2$ labeled Werner

are both compatible with the experimental facts but differ again in the mode of attachment of the 'reactive' chlorine atoms. Jørgensen regarded these atoms as linked to the metal atom through ammonia molecules, while Werner considered them as not linked to any particular atom, but attracted to the complex cation as a whole by electrostatic forces.

Werner explained the formation of the purpureo salt from the luteo salt by evolution of ammonia as a conversion of one of the three chlorine atoms from a primary (ionic) to a secondary (nonionic) valency. The entrance of the negative chlorine into the complex cation lowers the charge of the latter by one, and the charge of the resulting complex cation is now two, rather than three. Jørgensen criticized this interpretation, arguing that if a given negative group is coordinated to the central metal atom, it cannot simultaneously satisfy one of the primary valencies of the metal, a point which Werner later clarified. Werner's structure requires that a solution of purpureo cobaltic chloride furnish three ions, a fact confirmed by conductivity studies (Werner and Miolati, 1893, 1894). In further investigations, Jørgensen showed that the 'masked' chlorine atom in purpureo cobaltic salts could be replaced by other groups, such as bromide, sulfate, nitrate or oxalate. He also succeeded in preparing chloro-purpureo salts of chromium(III), completely analogous to those of cobalt(III), as well as similar chromium(III) salts in which the 'masked' chlorine had been replaced by bromine or iodine. In addition, he prepared a series of analogous purpureo rhodium(III) salts in which the 'masked' group is chlorine, bromine, iodine or nitrate.

(b) Aquapentaammines   (roseopentammines), $[M(NH_3)_5 H_2 O] X_3$. These compounds, so-called because of their rose-red color, are formed from purpureo salts by an aquation reaction:

$$[Co(NH_3)_5 Cl]Cl_2 + H_2O \rightleftharpoons [Co(NH_3)_5 H_2O]Cl_3$$

A molecule of water replaces the 'masked' chlorine atom, which now becomes ionic, whereupon all the chlorine atoms can be precipitated by silver nitrate. The reaction is reversible since at elevated temperatures water is lost, and the purpureo salt is regenerated.

Jørgensen (1884) regarded roseo compounds as luteo salts in which one-sixth of the ammonia had been replaced by water. Roseo salts are similar to luteo salts in formation, appearance, crystal structure, solubility and reactions. All the negative groups are ionic in both series of compounds. Both Jørgensen's and Werner's proposed structures

<pre>
       H₂O—Cl
      /
  Co—NH₃—Cl   [Co(NH₃)₅H₂O]Cl₃
      \
       NH₃—NH₃—NH₃—NH₃—Cl

       Jørgensen   Werner
</pre>

are compatible with the experimental facts. Each agreed that the water molecule was bonded to the metal atom, but they disagreed on the fate of the chlorine atom that had been replaced by the water molecule. Jørgensen regarded it as bonded to the cobalt atom through the oxygen of the water molecule, oxygen here being considered tetravalent just as nitrogen in ammonia chains was considered quinquevalent. Again, Werner considered it not bonded to any particular atom but attracted to the complex cation as a whole by electrostatic forces.

Jørgensen also succeeded in preparing analogous series of aquapenta-ammine salts of chromium(III) and rhodium(III).

(c)   Nitropentaammines   and   nitritopentaammines   (xantho   and isoxantho salts), $[M(NH_3)_5NO_2]X_2$ and $[M(NH_3)_5ONO]X_2$. By the action of a mixture of nitrogen(II) and (IV) oxides on an ammoniacal solution of cobalt(II) sulfate, Gibbs and Genth (1857) obtained yellow-brown prisms, $[Co(NH_3)_5NO_2]SO_4$, which they called xantho cobaltic sulfate. The corresponding chloride $[Co(NH_3)_5NO_2]Cl_2$ was obtained by metathesis with barium chloride. The NO_2 group in these compounds is unusually stable. It is not decomposed by acetic acid or even mineral acid solutions, in marked contrast to the behavior of the NO_2 group in nitrites. Jørgensen prepared a series of xantho rhodium(III) salts, ana-logous to those of cobalt(III) and chromium(III), but even more stable. He accounted for the stability of such compounds by assuming that the two nitrito groups are linked together through quinquevalent nitrogen:

$$Co_2 \begin{cases} -O-N{=}O \\ -O-\overset{\text{II}}{N}{=}O \\ NH_3-NH_3-NH_3-NH_3-Cl \\ NH_3-NH_3-NH_3-NH_3-Cl \\ NH_3-Cl \\ NH_3-Cl \end{cases}$$

When the introduction of monomeric formulae made this explanation unlikely and when he discovered an isomeric (isoxantho) series of salts, Jørgensen came to regard xantho salts as nitropentaammines rather than nitritopentaammines.

In 1894 Jørgensen prepared xantho cobaltic chloride by dissolving purpureo cobaltic chloride, $[Co(NH_3)_5Cl]Cl_2$, in dilute ammonia, neutralizing with hydrochloric acid, adding sodium nitrite, heating, and finally cooling and adding concentrated hydrochloric acid. When he merely let the solution stand in the cold after addition of the sodium nitrite, he obtained a new red compound isomeric with the xantho chloride. This compound, which he named isoxantho cobaltic chloride, is unstable toward acids and undergoes slow transformation to the xantho compound both in solution and in the solid state. The electrical con-ductivities of the xantho and isoxantho salts are almost identical and are indicative of the formation of three ions in solution (Werner and Miolati, 1894).

Jørgensen regarded the stable yellow-brown xantho salts as nitro ($-NO_2$) compounds because of the great stability of the Co$-$N bond and because the color of compounds containing six such bonds, such as luteo salts, ranges from yellow to brown. He regarded the unstable red iso-xantho salts as nitrito ($-O-N{=}O$) compounds because of the instability of the nitrito group toward acids and because compounds containing five Co$-$N bonds and one Co$-$O bond, such as roseopentammine salts, are red. On these points, Jørgensen and Werner were in perfect agreement, but as usual they differed on the general structure:

$$Co \overset{NO_2}{\underset{NH_3-NH_3-NH_3-NH_3-Cl}{\overset{NH_3-Cl}{\big\langle}}} \qquad Co \overset{O-N=O}{\underset{NH_3-NH_3-NH_3-NH_3-Cl}{\overset{NH_3-Cl}{\big\langle}}}$$

Xantho   Jørgensen   Isoxantho

$[Co(NH_3)_5NO_2]Cl_2$ $[Co(NH_3)_5O-N=O]Cl_2$

Werner

This type of isomerism, called structural isomerism, salt isomerism or linkage isomerism, is possible when more than one atom in a ligand can act as the donor to the central metal ion, i.e. when the ligand is ambidentate (Kauffman, 1973c). Although some workers have expressed doubt as to the existence of distinct xantho and isoxantho isomers, such isomerism has definitely been verified, and studies of the isomerization* reaction have been made. Thus the nitropentaammine and nitritopentaammine salts of cobalt(III) constitute not only the first and best known case of linkage isomerism among complexes but also the most extensively studied case. The xantho and isoxantho compounds remained the only confirmed case of linkage isomerism for many years until the corresponding nitritopentaammines of rhodium(III), iridium(III) and platinum(IV) were synthesized for the first time, and their isomerization to the nitropentaammines was studied (Basolo and Hammaker, 1962).

Type MA_4B_2 — tetraammines, $[M(NH_3)_4X_2^{-n}]Y_{3-2n}$
These compounds may be regarded as luteo salts in which one-third of the ammonia has been replaced by other groups. It is among such compounds that we first encounter the possibility of stereoisomerism, and therefore our detailed consideration of these compounds will be postponed until the section on configuration. Consequently, it suffices here to point out only that as far as constitution is concerned, both the Blomstrand—Jørgensen and Werner formulae for this type of compound agreed in the number of ions predicted, viz. two ions:

$$Co \overset{Cl}{\underset{NH_3-NH_3-NH_3-NH_3-Cl}{\overset{Cl}{\big\langle}}} \qquad [Co(NH_3)_4Cl_2]Cl$$

Jørgensen   Werner

and this fact was confirmed by conductivity studies (Werner and Miolati, 1893, 1894).

Type MA_3B_3 — triammines, $[M(NH_3)_3X_3]^0$

These compounds may be regarded as luteo salts in which one-half of the ammonia has been replaced by other groups. This type of compound played a most prominent role in the supersession of the Blomstrand—Jørgensen chain theory by the Werner coordination theory.

So far we have seen how Jørgensen's and Werner's formulations for hexaammines, pentaammines and tetraammines, both systems reasonably compatible with experimental facts, permitted two rival hypotheses to exist side by side for a limited time. However, the scientific mind feels uneasy at accepting two alternative explanations for a given group of phenomena, the coexistence of the wave and corpuscular theories of light notwithstanding. As more experimental evidence accumulated, the scales began to tip in favor of Werner's theory.

When successive ammonia molecules in a hexaammine are replaced by negative groups such as chlorine atoms, these enter the coordination sphere and thus become nonionic or 'masked'. With the replacement of the first two ammonia molecules, the ionic character of the compounds as predicted by the two theories is in complete agreement, but with that of the third ammonia molecule, the ionic character of the resulting compounds differs radically according to the two theories:

$$Co \diagup^{Cl} \!\!-Cl \diagdown_{NH_3-NH_3-NH_3-Cl} \qquad\qquad [Co(NH_3)_3Cl_3]^0$$

Two ions	Nonelectrolyte
Jørgensen	Werner

Jørgensen predicted that the chain of four ammonia molecules would merely be shortened by one and that the resulting compound would be similar to the preceding one in forming two ions in solution, one of the chlorine atoms still remaining ionic. On the other hand, Werner (1893) predicted an abrupt change in properties. The resulting compound should be a nonelectrolyte soluble in nonpolar solvents, and such solutions should not conduct an electric current. Werner pointed out that the properties of such compounds agreed with his theoretical predictions. Jørgensen (1894a) protested that the very few triammine complexes of trivalent metals then known were too poorly characterized to allow any conclusions to be drawn.

Werner prepared a blue-green compound $[Co(NH_3)_3Cl_3]$ whose solution does not readily yield a precipitate with silver nitrate, indicating the absence of ionic chlorine. In 1889 W. Palmaer had prepared the corresponding iridium compound $[Ir(NH_3)_3Cl_3]$, which did not evolve hydrogen chloride when heated with concentrated sulfuric acid, further evidence of the absence of ionic chlorine in compounds of this type.

Jørgensen himself prepared the rhodium compound $[Rh(NH_3)_3Cl_3]$ and found no hydrogen chloride evolution on sulfuric acid treatment. He tried to account for this conflicting evidence by attempting to discredit the sulfuric acid treatment as a test for ionic chlorine. He pointed out that Magnus' Green Salt, $[Pt(NH_3)_4][PtCl_4]$, is resistant to this treatment. Such reasoning seems fallacious since this compound does not contain ionic chlorine. Additional compounds of the MA_3B_3 type, which Jørgensen claimed are salts not nonelectrolytes, include $[Co(NH_3)_3 - (NO_3)_3]$ and $[Co(NH_3)_3(NO_2)_3]$. Since the last mentioned compound, trinitrotriamminecobalt(III), was, extremely crucial in the Werner– Jørgensen controversy, we shall now consider in detail the conductivity data for it and for the transition series $[Co(NH_3)_6](NO_2)_3 -$ $K_3[Co(NO_2)_6]$, of which it is a part.

Table IV shows a comparison of the formulae and predicted numbers of ions for this series according to the two theories. According to the Blomstrand–Jørgensen chain theory, the top compound, hexaamminecobalt(III) nitrite, to use modern IUPAC nomenclature, should dissociate to form three nitrite ions in solution because all three nitrite groups are bonded through the ammonia chains. Three nitrite ions plus the remainder of the compound, which forms a tripositive cation, results in the formation of a total of four ions. As we have seen, Jørgensen regarded four as the maximum number of ammonia molecules that can enter into a chain, which is therefore particularly stable, and his formulae contained a chain of four ammonia molecules whenever possible.

Loss of an ammonia molecule from the hexaammine results in formation of the second compound, nitropentaamminecobalt(III) nitrite, which according to the Blomstrand–Jørgensen formulation should furnish a total of only three ions – two nitrite ions, which are bonded through the ammonia chains, plus the remainder of the compound which functions as a dipositive cation. The uppermost nitrite group does not ionize because it is bonded directly to the cobalt atom. Loss of an ammonia molecule from the pentaammine results in the formation of the third compound, dinitrotetraamminecobalt(III) nitrite, which should furnish a total of only two ions – one nitrite ion, which is bonded through the ammonia chain, plus the remainder of the compound, which functions as a monopositive cation. The two uppermost nitrite groups do not ionize because they are bonded to the cobalt atom. Loss of an ammonia molecule from the tetraammine merely shortens the one remaining ammonia chain by one member, and therefore the resulting compound, trinitrotriamminecobalt(III), should form two ions in solution. Loss of further ammonia molecules would result in the formation of nonexistent compounds.

Now let us look at the formulae and predicted numbers of ions according to Werner's theory. According to Werner, as we have seen, loss of

TABLE IV
Constitution of cobalt(III) coordination compounds
(Kauffman, 1967c, 1976b, 1977c)

| Class of compound | Blomstrand—Jørgensen | |
	Formula	No. of ions
Hexaammines MA_6	$Co \begin{smallmatrix} NH_3-NO_2 \\ NH_3-NO_2 \\ NH_3-NH_3-NH_3-NH_3-NO_2 \end{smallmatrix}$ $\downarrow -NH_3$	4
Pentaammines MA_5B	$Co \begin{smallmatrix} NO_2 \\ NH_3-NO_2 \\ NH_3-NH_3-NH_3-NH_3-NO_2 \end{smallmatrix}$ $\downarrow -NH_3$	3
Tetraammines MA_4B_2	$Co \begin{smallmatrix} NO_2 \\ NO_2 \\ NH_3-NH_3-NH_3-NH_3-NO_2 \end{smallmatrix}$ $\downarrow -NH_3$	2
Triammines MA_3B_3	$Co \begin{smallmatrix} NO_2 \\ NO_2 \\ NH_3-NH_3-NH_3-NO_2 \end{smallmatrix}$	2
Diammines MA_2B_4	Unaccountable	—
Monoammines MAB_5	Unaccountable	—
Double Salts, MB_6	Unaccountable	—

ammonia molecules from ammines is not actually a simple loss, but rather a substitution in which a change in function of the anion occurs simultaneously, that is, as each molecule of ammonia leaves the coordination sphere, shown by square brackets, its place is taken by an anion which is no longer bonded by a primary or ionizable valence but instead by the secondary or nonionizable valence vacated by the departing ammonia molecule. The charge of a complex should be equal to the algebraic sum of the charges of the central metal ion and of the coordinated groups. Consequently, as neutral molecules of ammonia (A) in a metal–ammine (MA_6) are successively replaced by anions (B), the number of ions in the resulting compounds should progressively decrease until a nonelectrolyte

Werner	
Formula	No. of ions
$[Co(NH_3)_6](NO_2)_3$	4
$\downarrow -NH_3$	
$[Co(NH_3)_5NO_2](NO_2)_2$	3
$\downarrow -NH_3$	
$[Co(NH_3)_4(NO_2)_2]NO_2$	2
$\downarrow -NH_3$	
$[Co(NH_3)_3(NO_2)_3]$	0
$\downarrow -NH_3 + KNO_2$	
$K[Co(NH_3)_2(NO_2)_4]$	2
$\downarrow -NH_3 + KNO_2$	
Unknown for Cobalt	(3)
$\downarrow -NH_3 + KNO_2$	
$K_3[Co(NO_2)_6]$	4

is formed and then increase as the complex becomes anionic. How do the predictions of the two theories agree with the experimental facts? Let us examine the conductivity data in some detail.

Werner's first published experimental work in support of his coordination theory was a study of conductivities carried out during the years 1893–1896 in collaboration with his friend and former fellow student Arturo Miolati (1869–1956), who was later to become Professor of Chemistry at the Universities of Turin and Padua (Kauffman, 1970a). The two young collaborators are shown arm in arm in Plate 16.

The chemistry of complexes had long been intimately linked with electrochemistry. In fact, the discovery of complex ions *per se* has usually

Plate 16. Arturo Miolati (left) and Alfred Werner (right) [Courtesy, Fräulein Crescentia R. Roder, Bonn, German Federal Republic]

been attributed to Johann Wilhelm Hittorf (1824–1914), who in his classical series of studies on ionic migration (1853–1859) established the fundamental distinction between double salts* and complex salts*. In 1875, Friedrich Kohlrausch (1840–1910) discovered experimentally the law of independent migration of ions, and four years later he established the principle that the equivalent conductivity of an electrolyte can be represented as the sum of the conductivity due to the cation and of that due to the anion. Inasmuch as ionic mobilities do not vary greatly from one ion to another, except for H^+ and OH^-, Kohlrausch's additivity principle provided a relatively simple way to determine the number of ions in a soluble compound. It was this physicochemical method that Werner and Miolati applied to the solution of an important chemical problem – the constitution of complexes (Kauffman, 1968).

In their first joint publication on this subject (Werner and Miolati, 1893), thy showed that the molecular conductivities (μ) of coordination compounds decrease as successive molecules of ammonia are replaced by acid residues (negative groups or anions). For example, in the case of cobalt(III) salts, they found that μ for luteo salts (hexaammines) $> \mu$ for purpureo salts (acidopentaammines) $> \mu$ for praseo salts (diacidotetraammines). The conductivity falls almost to zero for the triacidotriammine

$Co(NH_3)_3(NO_2)_3$, and then rises again for tetraacidodiammines, in which the complex behaves as an anion. On the other hand, substitution of an ammonia molecule by a water molecule produces little change in conductivity.

By such conductivity measurements, Werner and Miolati were able to determine the number of ions in complexes of cobalt(III) and platinum(II) and (IV). In this way, they not only found support for the coordination theory, but they also elucidated the process of dissociation of salts in aqueous solution. Moreover, they were able to follow the progress of aquation reactions by measuring changes in conductivity with time.

The data obtained for $Co(NH_3)_3(NO_2)_3$ were of paramount importance in deciding between Werner's and Jørgensen's views. In Jørgensen's own words (1894b), 'This is the central point in Werner's system. With this [point], it stands or falls'. The Blomstrand—Jørgensen theory predicted that such a compound would be an electrolyte, whereas, in direct opposition to this view, Werner had stated: 'In compounds $M(NH_3)_3X_3$, absolutely no negative complex continues to exhibit the behavior of an ion' (Werner, 1893). Jørgensen, on the other hand, insisted that 'the process of formation of $M(NH_3)_3X_3$ compounds [from tetraammines] is completely different from that of the formation of purpureo salts [pentaammines] from luteo salts [hexaammines] . . . . Loss of an ammonia [molecule] produces no change in the chemical function of an acid residue'. The correctness of Werner's views was proven by his and Miolati's conductivity values.

In the second of their articles on the conductivities of complexes, Werner and Miolati (1894) demonstrated the complete agreement in magnitude, variation and pattern between their experimentally measured conductivities and those predicted according to the coordination theory. They compiled values for the conductivities of cobalt(III), chromium(III) and platinum(II) and (IV) complexes containing different numbers of ions. They found that, for a given type of compound, conductivity values are of the same order of magnitude and fluctuate only within narrow limits so that a conductivity measurement can be used to establish to which type of compound an unknown complex belongs.

Every book dealing with coordination chemistry, whether introductory or advanced, includes almost without exception the familiar V-shaped plots of the molecular conductivities of cobalt(III)— and platinum(II)— and (IV)—ammines vs the number of chlorine atoms introduced into the coordination sphere. The popularity of these graphs can probably be attributed to the numerous fundamental postulates of Werner's theory that they illustrate — the formation of a transition series from ammines to double salts by the stepwise displacement of a neutral ligand by an

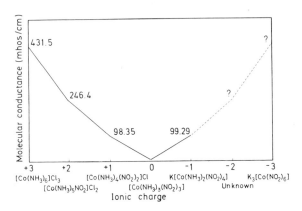

Figure 1. Conductivities of cobalt(III) coordination compounds
(Kauffman, 1967c, Werner and Miolati, 1894).

anion entering the coordination sphere, the effect of such entry on the
charge of the complex and on the number of ions present, the determination of the number of ions by conductivity measurements and the main-
tenance of a constant coordination number during these changes. The
graph of experimentally measured conductivities for the transition series
$[Co(NH_3)_6]Cl_3 - K_3[Co(NO_2)_6]$ (Fig. 1) agreed completely with the
numbers of ions predicted according to the coordination theory (Table
IV). The conductivity results were also concordant with the number of
'masked' and 'unmasked' acid radicals.

In their second article too, Werner and Miolati attempted to answer
some of Jørgensen's objections to their first article. On the whole,
Jørgensen (1894b) regarded their experimental work favorably and considered their conductivity data as confirming his views on the constitution
of luteo, pentammine- and tetrammineroseo and pentammine- and
tetramminepurpureo salts as well as proving his own observations concerning the easy transition of many purpureo salts into roseo salts. This is not
unexpected since the number of ions predicted for these compounds is
the same for both the Werner theory and the Blomstrand–Jørgensen
theory.

Jørgensen, however, attempted to discredit the conductivity data
obtained for Erdmann's triamminecobalt nitrite, $Co(NH_3)_3(NO_2)_3$,
inasmuch as neither he nor Wolcott Gibbs (1822–1908) had been able to
prepare this crucial compound according to Erdmann's directions.
Furthermore, he regarded Werner and Miolati's proof of its identity solely
on the basis of a cobalt determination as inconclusive since he himself had
prepared eight double salts which were isomeric with the triamminine
(1894a). He went on to speculate that Werner and Miolati might actually

have prepared one of these double salts and that perhaps such a compound would exhibit minimal conductivity.

Werner and Miolati (1894) tried to measure the conductivities of all eight of Jørgensen's isomeric compounds, but, because of limited solubilities, they were able to accomplish this for only four of them. They also made a detailed study of the conductivity of dichrocobalt chloride inasmuch as Jørgensen (1894), in his attempts to establish the fact that $Co(NH_3)_3(NO_2)_3$ is an electrolyte, had carefully examined the closely related dichroic compound. Although Jørgensen formulated this salt as

$$Co \begin{array}{l} \diagup OH_2 \cdot Cl \\ -NH_3 \cdot Cl \\ \diagdown NH_3 \cdot NH_3 \cdot Cl \end{array}$$

with three ionizable chlorine atoms, Werner and Miolati showed it to be $[Co(NH_3)_3(H_2O)Cl_2]Cl$, which, however, by an aquation reaction eventually forms $[Co(NH_3)_3(H_2O)_3]Cl_3$. Not only did this effectively parry Jørgensen's thrust aimed at $Co(NH_3)_3(NO_2)_3$, but it also underscored the value of conductivity measurements as a technique for following the course of chemical reactions.

In short, the classical conductivity studies of Werner and Miolati on a wide variety of metal–ammine complexes agreed very well with Werner's theory. The conductivities of compounds such as $[Co(NH_3)_3(NO_2)_3]$ were found to be extremely low, an indication of nonelectrolytic character. Petersen (1897) verified Werner and Miolati's experiments but objected to their conclusions in those cases where the conductivities corresponded to a greater number of ions than that predicted by the coordination theory. Werner and Miolati explained these apparent discrepancies by aquation reactions such as:

$$[Co(NH_3)_3(NO_2)_2Cl]^0 + H_2O \rightarrow [Co(NH_3)_3(NO_2)_2(H_2O)]^+ + Cl^-$$

However, measurements of compounds not containing as ligands groups readily displaced by water agreed completely with the theory. Petersen also tried to verify his conductivity measurements by cryoscopic studies, but he encountered some discrepancies. Jørgensen seized upon these so-called 'discrepancies' in an attempt to discredit the entire conductivity method and hence all of Werner and Miolati's results. Actually, Petersen's results did not support Jørgensen's view any better than they did Werner's theory.

Furthermore, in its explanation of anionic complexes and its demonstration of the existence of a continuous transition series (*Übergangsreihe*) between metal–ammines (MA_6) and double salts (MB_6), the Werner theory succeeded in an area in which the Blomstrand–Jørgensen theory could not pretend to compete.

Until 1907, when Werner discovered the long-sought ammonia–violeo salts (*cis*-dichlorotetraammines) (see pp. 116–119) Jørgensen maintained his steadfast conviction that Blomstrand's theory was the only one compatible with the experimental facts. Although he lived until 1914, he discontinued his polemic with Werner in 1899. His final plea on behalf of Blomstrand's views, written at the age of 62, marks his gracious withdrawal from the controversy:

> As far as the metal–ammonia salts are concerned, I have now compared the older theory with Werner's system from all essential points of view and as objectively as I could, and I am now even more convinced than previously that Blomstrand's theory together with my modifications still explains best all actual relationships in the chemistry of the metal–ammonia salts in accordance with our whole chemical system. It was my task to examine this question exactly as it relates to Werner's theory. For he is the only one who has tried to build a new theory with the consideration of all details. He has not been satisfied with casual suggestions, which are of absolutely no use if one wishes to apply them to the compounds discussed in this and previous works. Furthermore, it is to his credit that not only has he stimulated new researches by his many striking and ingenious observations, but he himself has revealed many new, interesting, and important facts, and I hope that he will reveal many more new facts (Jørgensen, 1899b; Kauffman, 1976b).

It was only natural that Werner's views, marking a sharp break in the classical theory of valency and structure, should have seemed too radical to Jørgensen. When Werner first proposed his theory, the octahedral configuration for cobalt(III) was an *ad hoc* explanation, a mere guess without adequate experimental verification. At this time, on the other hand, the older man had already devoted many years to thorough investigations of metal–ammine complexes and had accounted for his findings by a consistent application of Blomstrand's chain theory, which he modified, but only when absolutely necessary.

Thus the controversy between Jørgensen and Werner over the constitution of metal–ammine complexes provides us with an excellent illustration of the synergism so often encountered in the history of science. During the course of this competition, conducted without any trace of jealousy or rancor, each chemist did his utmost to prove his views, and in the process a tremendous amount of fine experimental work was performed by both. Although not all of Jørgensen's criticisms were valid, Werner, in many cases, was forced to modify various aspects of his theory. However, the basic postulates were verified in virtually every particular.

Werner's ideas eventually triumphed, but Jørgensen's experimental observations are thereby in no way invalidated. On the contrary, his experiments, performed with extreme care, have proved completely reliable and form the foundation not only of the Blomstrand—Jørgensen theory but of Werner's as well. From the very beginning of the controversy, Werner continually acknowledged his great debt to the older man. For example, in 1913, on his way to Stockholm to receive the Nobel prize, Werner addressed the Danish *Kemisk Forening* in Copenhagen, acknowledging the important role that Jørgensen's experimental contributions had played in the development of the coordination theory.

CONFIGURATION OF COBALT—AMMINES

Now let us examine the means used by Werner to establish the configuration of cobalt—ammines (Kauffman 1977c). The technique of 'isomer counting' that he used as a means of proving configuration admittedly did not originate with Werner. The idea of an octahedral configuration and its geometric consequences with respect to the number of isomers expected had been considered as early as 1875 by Jacobus Henricus van't Hoff, and the general method is probably most familiar through Wilhelm Körner's work of 1874 on disubstituted and trisubstituted benzene derivatives. Yet the technique of comparing the number and type of isomers actually prepared with the number and type theoretically predicted for various configurations probably reached the height of its development with Werner's work. By this method, he was able not only

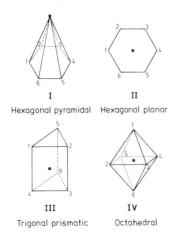

Figure 2. Configurational possibilities for coordination number 6.

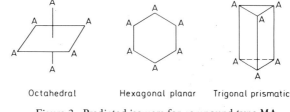

Figure 3. Predicted isomers for compound type MA_6.

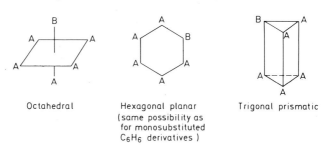

Figure 4. Predicted isomers for compound type MA_5B.

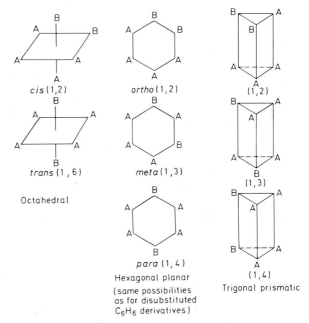

Figure 5. Predicted isomers for compound type MA_4B_2.

to discredit completely the rival Blomstrand—Jørgensen chain theory but also to demonstrate unequivocally that trivalent cobalt possesses an octahedral condiguration rather than another possible symmetrical arrangement such as hexagonal planar or trigonal prismatic.

The method is indirect but basically simple. For coordination number six, if all six positions are equivalent, four configurations are possible — hexagonal pyramidal, hexagonal planar, trigonal prismatic and octahedral (Fig. 2), of which only the last three are usually considered. Table V shows the predicted number of isomers theoretically possible for selected compound types according to each of the three different configurations. For compound type MA_6 or hexaammines, each of the three configurations should result in the same number of geometric isomers, namely one, as shown in Fig. 3. For compound type $MA_5 B$ or pentaammines, each of the three configurations should also result in the same number of geometric isomers, namely one, as shown in Fig. 4. Therefore the number of isomers actually found in these two cases does not permit a choice between the three configurations. On the other hand, as shown in Fig. 5, in the case of compound type $MA_4 B_2$ or tetraammines, the hexagonal

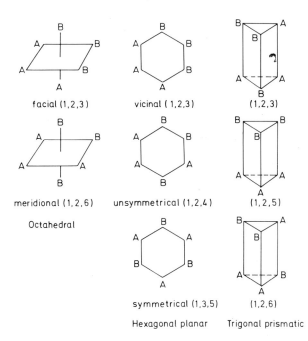

Figure 6. Predicted isomers for compound type MA_3B_3.

planar and trigonal prismatic configurations would each result in three possible geometric isomers, whereas the octahedral configuration would result in only two possible geometric isomers. Compounds of types such as MA_4B_2, MA_4BC, $M(\overline{AA})_2B_2$, or $M(\overline{AA})_2BC$ should exist in two forms for an octahedral configuration but in three forms for the other two configurations, and consequently most of Werner's synthetic attempts involved compounds of these types (Werner, 1912). As shown in Fig. 6, for compound type MA_3B_3 or triammines, again the hexagonal planar and trigonal prismatic configurations would each result in three possible geometric isomers, whereas the octahedral configuration would result in only two possible geometric isomers. As shown in Fig. 7, for compound type $M(\overline{AA})_3$, that is, trisbidentate complexes or those containing three bidentate chelate groups, each of the three possible configurations would result in different isomeric possibilities, and these will be examined in more detail later (pp. 121–136). In most cases, as a comparison of columns IV and V of Table V shows, the number and type of isomers actually prepared corresponded to the theoretical expectations for the octahedral arrangement, but there were a few exceptions, and Werner required more than two decades to accumulate a definitive proof for his structural ideas.

In considering Werner's proof of the configuration of the cobalt–ammines we shall examine in detail two types of stereoisomerism, viz. geometric isomerism and optical isomerism.

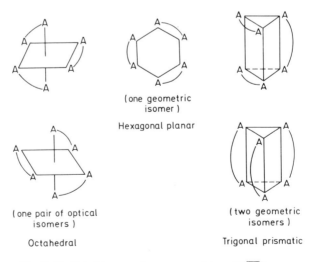

Fig. 7. Predicted isomers for compound type $M(\overline{AA})_3$.

TABLE V
Proof of configuration for coordination number six by 'isomer counting' (Kauffman, 1976b)

| Compound type | Theoretically predicted isomers | | | | Experimentally found isomers | Result VI |
	I Hexagonal pyramidal	II Hexagonal planar (a special case of I)	III Trigonal prismatic[b]	IV Octahedral	V	
MA_6	One form only	One form only	One form only	One form only	One form only	None
MA_5B	One form only	One form only	One form only	One form only	One form only	None
$M(AA)B_4$[a]	One form only	One form only	Two geometric	One form only	One form only	Provisionally eliminates trigonal prismatic (III)
MA_4B_2	Three geometric (1,2; 1,3; 1,4)	Three geometric (1,2; 1,3; 1,4)	Three geometric (1,2; 1,3; 1,4)	Two geometric (1,2 cis; 1,6 trans)	Two or less geometric	Provisionally proves octahedral (IV); discovered 1907
MA_3B_3	Three geometric (1,2,3; 1,2,4; 1,3,5)	Three geometric (1,2,3; 1,2,4; 1,3,5)	Three geometric (1,2,3; 1,2,5; (1,2,6)	Two geometric (1,2,3 facial; 1,2,6 meridional)	Two or less geometric	Provisionally proves octahedral (IV)
$M(\overline{AA})_2B_2$[a] or $M(AA)_2BC$[a]	Two geometric	Two geometric	Four geometric, one of which is asymmetric	Two geometric (1,2 cis; 1,6 trans), the first of which is asymmetric	Two geometric (1,2 cis; 1,6 trans), the first of which was resolved	Unequivocally proves octahedral (IV); discovered 1911
$M(\overline{AA})_3$[a]	One form only	One form only	Two geometric	One asymmetric pair	One pair optical resolved	Unequivocally proves octahedral (IV); discovered 1912

[a] $\overline{AA}$ represents a symmetrical bidentate (chelate) ligand. Such ligands coordinate at two adjacent positions. They can span cis positions but not trans positions.

[b] Coordination compounds with this configuration have now been synthesized (R. Eisenberg and J. A. Ibers, J. Am. Chem. Soc. 87, 3776–3778 (1965); Inorg. Chem. 5, 411–416 (1966); E. I. Stiefel and H. B. Gray, J. Am. Chem. Soc. 87, 4012–4013 (1965); H. B. Gray, R. Eisenberg and E. I. Stiefel, in Werner Centennial, G. B. Kauffman, Symposium Chairman, American Chemical Society, Washington, DC, 1967, pp. 641–650).

Geometric isomerism (Kauffman, 1959, 1960, 1967c, 1968, 1975a, 1977c)
From his earliest work on coordination compounds, Werner recognized that, contrary to common belief, isomerism was an important and frequently encountered phenomenon in inorganic chemistry and that it could provide insight into the structure of complex compounds. His remarks on pp. 158–159 of the first edition of his *Neuere Anschauungen* are typical:

> 'The concept of isomerism has only minor significance in the inorganic field, especially since only a few cases of isomerism are known', is an assertion which is repeatedly found in the most diverse variations in the publications of the last decade and which is occasionally still found repeated today without any proof for its justification . . . . The compounds in the inorganic field, with the exception of the simple bases, acids and salts, are of very complex composition; furthermore, they are synthetically either very difficult to obtain or their mode of formation gives no information about their constitution. It is therefore important that the appearance of important isomeric phenomena gives us the means of fathoming the involved structure of these complicated compounds.
>
> That the concept of isomerism actually has great significance in inorganic chemistry also and that the number of isomeric phenomena is no small one will be seen from the following. First, it should be emphasized that cases of inorganic isomerism are more varied than organic ones, since they do not originate from a unified principle but rather must be attributed to varied and in part still not wholly recognized causes (Werner, 1905).

By 1912, in the opening lines of his longest paper Werner was able at long last to announce triumphantly:

> The investigations of stereoisomeric cobalt compounds have occupied us longer than we originally intended, primarily because for a long time it was not possible to discover conclusive evidence for the steric concept and unambiguous methods for the determination of configuration. It was therefore necessary to accumulate a very extensive collection of factual data before positive experimental foundations for the solution of these problems could be attained. This has now been accomplished, and the experimental results published in this article demonstrate that both problems have finally been solved (Werne, 1912).

Through the years Werner and his students succeeded in preparing and characterizing, in most cases for the first time, geometric isomers of a wide variety of cobalt complexes (Werner, 1912; Kauffman, 1975a). Among trivalent complexes he isolated isomeric salts of the following series: $[Co(en)_2(NH_3)_2]X_3$, $[Co(en)_2(H_2O)_2]X_3$, $[Co(en)_2(NH_3)-H_2O]X_3$, $[Co(NH_3)_4(H_2O)_2]X_3$ and $[Co(NH_3)_3(H_2O)_3]X_3$. Among divalent complexes he isolated isomeric salts of the following series: $[Co(en)_2(NH_3)NO_2]X_2$, $[Co(en)_2(NH_3)NO_3]X_2$, $[Co(en)_2(NH_3)F]X_2$, $[Co(en)_2(NH_3)Cl]X_2$, $[Co(en)_2(NH_3)Br]X_2$, $[Co(en)_2(NH_3)NCS]X_2$, $[Co(en)_2(H_2O)OH]X_2$, $[Co(en)_2(H_2O)NO_2]X_2$, $[Co(en)_2(H_2O)Cl]X_2$, $[Co(en)_2(H_2O)NCS]X_2$, $[Co(NH_3)_4(H_2O)OH]X_2$, $[Co(NH_3)_4(H_2O)-NO_2]X_2$, $[Co(NH_3)_4(H_2O)Cl]X_2$ and $[Co(NH_3)_3(H_2O)_2Cl]X_2$. Among monovalent complexes he isolated isomeric salts of the following series: $[Co(en)_2(N_3)_2]X$, $[Co(en)_2(NO_2)_2]X$, $[Co(en)_2(ONO)_2]X$, $[Co(en)_2-F_2]X$, $[Co(en)_2Cl_2]X$, $[Co(en)_2Br_2]X$, $[Co(en)_2(NCS)_2]X$, $[Co(en)-(pn)(NO_2)_2]X$ (pn = propylenediamine), $[Co(pn)_2(NO_2)_2]X$, $[Co(pn)_2-Cl_2]X$, $[Co(en)_2(OH)Cl]X$, $[Co(en)_2(OH)NCS]X$, $[Co(en)_2(NO_2)Cl]X$, $[Co(en)_2(NO_2)NCS]X$, $[Co(en)_2Cl(NCS)]X$, $[Co(en)_2Br(NCS)]X$, $[Co-(en)_2ClBr]X$, $[Co(en)(NH_3)_2Cl_2]X$, $[Co(NH_3)_4(NO_2)_2]X$, $[Co(NH_3)_4-Cl_2]X$, $M[Co(en)_2(SO_3)_2]$, $M[Co(NH_3)_4(SO_3)_2]$, $[Co(NH_3)_4(NO_2)-NCS]X$, $[Co(NH_3)_3(H_2O)Cl_2]X$ and $[Co(NH_3)_2(H_2O)_2Cl_2]X$. Among nonelectrolytes he isolated $[Co(NH_3)_3(NO_2)_3]$, $[Co(NH_3)_3(C_2O_4)NO_2]$ and $[Co(NH_3)_3(C_2O_4)Cl]$. Werner's work was not limited to the compounds of cobalt, for he also isolated the following isomeric chromium complexes: $[Cr(NH_3)_2H_2O(SCN)_3]H_2O$, $M[Cr(H_2O)_2(C_2O_4)_2]$, $M_2[Cr(H_2O)OH(C_2O_4)_2]$, $M_2[Cr(H_2O)OCOCH_3(C_2O_4)_2]$, $M_3[Cr-(OH)_2(C_2O_4)_2]$ and $M_3[Cr(OCOCH_3)_2(C_2O_4)_2]$.
A detailed treatment of all these geometric isomers would go far beyond the scope of this book, and we shall consider only three of the most important and well-known cases, all belonging to the class of tetraammines.

Type MA_4B_2 – tetraammines. (a) *cis-* and *trans-*Dichlorobis(ethylenediamine) salts (violeo and praseo salts), $[Co(en)_2Cl_2]X$. The first and still probably the best known case of geometric isomerism among inorganic complexes was discovered in 1890 by Jørgensen not among simple tetraammines MA_4B_2 but among salts of the $M(\overline{AA})_2B_2$ type, in which the four ammonia molecules have been replaced by two molecules of the bidentate (chelate) organic base ethylenediamine. Jørgensen first described green salts of composition $[Co(en)_2Cl_2]X$, completely analogous to the previously known praseo salts of Gibbs and Genth, $[Co(NH_3)_4Cl_2]X$, which he therefore called ethylenediaminedichloropraseo salts (Jørgensen, 1889). In the following year Jørgensen observed that repeated evaporation of a neutral aqueous solution of ethylenediaminedichloropraseo chloride produced an isomeric violet compound which he called ethylene-diaminedichlorovioleo chloride (Jørgensen, 1890a).

The two chlorine atoms in both series of $[Co(en)_2 Cl_2]X$ salts are 'masked', i.e. they do not yield a precipitate of silver chloride in the cold on treatment with silver nitrate solution. By cryoscopic* and conductivity measurements, Werner and his American student Charles Holmes Herty (1901) showed that their solutions contained a unipositive cation. Like the dichlorotetraammines, the dichlorobis(ethylenediamine) salts undergo aquation in solution, as shown by changes in color, in absorption spectra, in molecular conductivity and in flocculating power. Therefore, on stand-ing in solution, the two 'masked' chlorine atoms become ionic. This 'hydrolysis'* (actually, an aquation), which has been studied by numerous workers, is dependent upon the nature of the anions present and occurs in two steps:

(1) cis- or trans-$[Co(en)_2 Cl_2]^+ + H_2 O \rightarrow$ cis-$[Co(en)_2 (H_2 O)Cl]^{2+} + Cl^-$
(a slow reaction, which, however, is more rapid for the cis- than for the trans-dichloro compound) and
(2) cis-$[Co(en)_2 (H_2 O)Cl]^{2+} + H_2 O \rightarrow$ cis-$[Co(en)_2 (H_2 O)_2]^{3+} + Cl^-$

Aside from their color, the cis- and trans-$[Co(en)_2 Cl_2]X$ compounds differ in a number of properties such as absorption spectra and ionic mobility. Like many other trans salts, these trans compounds tend to form acid salts; e.g. in addition to Jørgensen's trans-$[Co(en)_2 Cl_2]Cl\cdot$HCl·$2H_2 O$ and Werner's acid salt, formulated as trans-$[Co(en)_2 Cl(HCl)\-(H_2 O)_2]Cl_2$, an entire series of acid salts with dibasic organic acids has been prepared. Trans-$[Co(en)_2 Cl_2]X$ salts react immediately with con-centrated aqueous NH_3 to give trans-$[Co(en)_2 (NH_3)Cl]X_2$ salts, while the cis compounds dissolve only on heating with formation of trans-$[Co(en)_2 (H_2 O)OH]X_2$ salts. The cis-dichlorobis(ethylenediamine) salts are converted into the trans salts on heating in acid solution, while the reverse reaction occurs on evaporating neutral solutions of the trans salts. Numerous studies have been made of the isomerization* reaction as well as of substitution reactions.

In his first paper on the coordination theory (1893), Werner, in dis-cussing the praseo and violeo salts, stated 'This interesting isomerism is the first confirmation of the conclusions resulting from the octahedral formula'. He regarded these compounds as cis and trans stereoisomers, that is, he considered them to contain the same atoms and bonds and to differ only in the orientation of these atoms and bonds in space. In other words, he felt that the isomerism was merely a geometric con-sequence of the octahedral structure.

Jørgensen, on the other hand, disagreed with this view, formulated the
compounds as Co·en·en·X $\begin{smallmatrix}\cdot Cl \\ \cdot Cl\end{smallmatrix}$ and considered the difference in color as due
to structural isomerism connected with the linking of the two ethylene-
diamine molecules, a situation that could not occur among the simple
tetraammines:

> Whereas the diatomic group $(NH_3)_4$ can be conceived as com-
> posed in only one manner, $-NH_3 \cdot NH_3 \cdot NH_3 \cdot NH_3-$, the di-
> atomic group $(NH_2 C_2 H_4 NH_2)_2$ can be conceived as composed
> in two ways, namely:

> (Jørgensen 1895). . . . only one conclusion seems possible,
> namely that such a difference between praseo and violeo salts
> as we find among the cobalt ethylenediamine salts does not
> occur among the cobalt ammonia salts (Jørgensen, 1897).

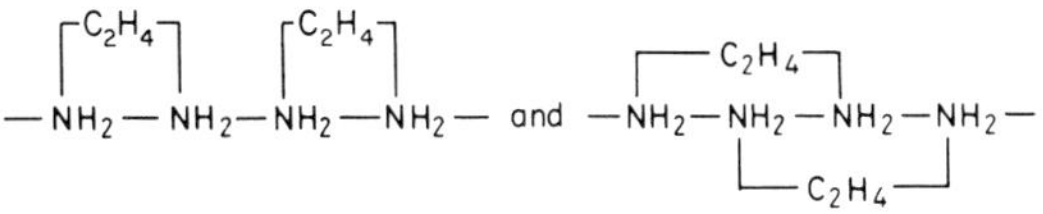

Jørgensen's and Werner's formulae for the ethylenediamine violeo and
praseo salts are contrasted in Fig. 8.

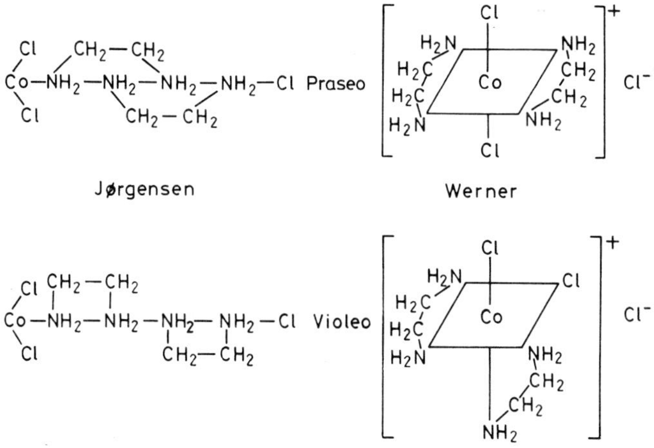

Figure 8. Jørgensen's and Werner's formulae for praseo and violeo
ethylenediamine isomers.

This may be an appropriate point at which to consider in general how Werner assigned configurations to isomers. From the inception of his coordination theory, he was concerned with the problem of assigning configurations to each isomer of given isomer pairs for both coordination numbers four and six. In other words, he devised methods for deciding which isomer was *cis* and which was *trans*. Basically, the methods devised were similar to those used in organic chemistry to assign structures to geometrically isomeric ethylene compounds. Both methods depend on the determination of the genetic relationships between one of the isomers and the corresponding cyclic compounds (Werner, 1912).

The compound for which the *cis* configuration is to be proven should be preparable from a closed ring compound or should be convertible into a compound of a known closed ring constitution. The simplest organic example is the case of fumaric acid:

$$\begin{array}{c} H-C-COOH \\ \| \\ HOOC-C-H \end{array}$$

and maleic acid:

$$\begin{array}{c} H-C-COOH \\ \| \\ H-C-COOH \end{array}$$

The latter is considered to be the *cis* form because it forms a cyclic anhydride, which on hydrolysis regenerates maleic acid:

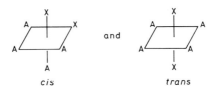

Similarly, among stereoisomeric cobalt compounds

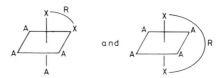

it is obvious that, since chelate groups can only span adjacent positions on the octahedron, ring closure between the two X atoms can occur easily for the *cis* isomer but only with great difficulty, if at all, for the *trans* isomer:

The strict analogy with the ethylene compounds can be made more obvious if we compare with the organic compounds the relative positions of the groups in the octahedral plane in which two A groups and two X groups are found:

cis   trans

Maleic acid   Fumaric acid

Werner assumed that in the carbonatotetraammine salts $[Co(NH_3)_4-CO_3]X$, the carbonate group forms a 4-membered ring with the cobalt atom

and that consequently the two oxygen atoms bonded to the cobalt atom are in the *cis* positions. He therefore concluded that when diaquatetraammine salts are formed from the carbonato compound by the action of dilute mineral acids, the two water molecules enter the *cis* positions:

$$[Co(NH_3)_4CO_3]X + 2HX + H_2O \rightarrow cis-[Co(NH_3)_4(H_2O)_2]X_3 + CO_2\uparrow$$

Since similar replacements among stereoisomeric cobalt complexes are known in many cases to be accompanied by rearrangement, Werner's choice of this reaction as a starting point for configurational determinations was an extremely fortunate one. In some cases, Werner confirmed these configurational assignments by resolution experiments – *cis* isomers were resolved, while all efforts to resolve *trans* isomers resulted in failure, as would be expected from the octahedral configuration (see p. 39). Although other, more dependable, methods for determining configuration of isomers have since been developed, Werner's assignments by classical chemical methods have in most cases withstood the test of time.

In the case of the dichlorobis(ethylenediamine)cobalt(III) isomers, Werner assigned the violeo salts the *cis* configuration by preparing them by treating with concentrated hydrochloric acid cobalt complexes containing closed rings such as:

$$\left[(en)_2 Co \diagup\!\!\!\diagup_O^O \diagdown\!\!\!\diagdown C=O \right] X \qquad \left[(en)_2 Co \diagup\!\!\!\diagup_O^O \diagdown\!\!\!\diagdown S=O \right] X$$

$$\left[(en)_2 Co \diagup\!\!\!\diagup_{HO}^{OH} \diagdown\!\!\!\diagdown Co(en)_2 \right] X_4$$

He confirmed this assignment of configuration by resolving them by means of ammonium (+)- and (−)-bromocamphor-π-sulfonates (Werner, 1911b). The resolution has been repeated by Bailar and Auten (1934), and subsequent resolutions by means of (+)- and (−)-quartz have been reported. The resolution of optical isomers and its relation to structure-proof will be dealt with later in this chapter (pp. 121−136).

(b) *cis*- and *trans*-Dichlorotetraammine salts (ammonia-violeo and praseo salts), [Co(NH$_3$)$_4$Cl$_2$]X. According to Werner, the isomerism among the [Co(en)$_2$Cl$_2$]X salts was merely a geometric consequence of the octahedral structure and should be observed in compounds of type MA$_4$B$_2$ that do not contain ethylenediamine. Thus his theory would predict the existence of a series of violeo (*cis*-dichlorotetraammine-cobalt(III)) salts, [Co(NH$_3$)$_4$Cl$_2$]X, isomeric with the corresponding praseo (*trans*) compounds. However, for these compounds, only one series (praseo, green) was known. Replacement of both nitro groups in flavo (*cis*-dinitrotetraamminecobalt(III)) salts with chlorine using dilute hydrochloric acid always resulted in formation of the praseo (*trans*) compounds, as did replacement of the carbonato group in carbonatotetra-amminecobalt(III) salts in the same manner. Thus the missing violeo salts, like the constitution of trinitrotriamminecobalt(III) before them (pp. 100−103), became another major point of contention between the two adversaries. Naturally, Jørgensen, being a confirmed empiricist, quite correctly felt justified in criticizing Werner's theory on the grounds that it predicted the existence of many compounds that were then unknown. Although the simple violeo salts were the most famous of such 'non-existent' compounds, they were by no means the only ones; Jørgensen (1897) listed more than a dozen series of such compounds.

Among the simple tetraammines not containing ethylenediamine, only NO$_2$ and SO$_3$ isomers were known, and it could be argued that such compounds were not stereoisomers but rather structural isomers caused by the difference in the linking of the NO$_2$ and SO$_3$ groups. Until 1906, all cases of stereoisomerism among MA$_4$B$_2$ type compounds had involved acid radicals (negative groups) for either A or B, which, according to Werner, were bound to the central metal ion by primary valence bonds (*Haupt-valenzen*), while the remaining groups were bound by secondary valence bonds (*Nebenvalenzen*). In that year, however, Werner *et al.* (1906) were able to announce the preparation of MA_4B_2 stereoisomers of formula $[Co(en)_2(NH_3)_2]X_3$, in which all ligands were bound to the central metal ion by secondary valence bonds, thus demonstrating that the occurrence of stereoisomerism is independent of whether or not acid radicals are bound directly to the central metal ion. But the missing violeo salts remained unknown.

In general, ammonia complexes are less stable than the corresponding ethylenediamine complexes. The preparation of the violeo compounds therefore posed a difficult and challenging synthetic problem, for they rapidly undergo aquation to form chloroaqua salts unless the temperature is kept low and the hydrochloric acid concentration is kept high. Under the circumstances, it is understandable that immediately after his successful preparation of these elusive compounds, even before submitting the manuscript to Paul Jacobson, editor of the *Berichte der Deutschen Chemischen Gesellschaft*, Werner (in a letter of 13 November 1907) jubilantly informed Jørgensen of his discovery:

> I am taking the liberty of sending you in the same mail a sample of the long-sought ammonia-violeo series $[Cl_2Co(NH_3)_4]Cl$ and hope that you too will take pleasure in it.

When Jørgensen learned of the preparation of these compounds, whose existence was a necessary consequence of the coordination theory but not of the Blomstrand—Jørgensen chain theory, he promptly acknowledged the validity of Werner's views.

The first dichlorotetraamminecobalt(III) salt to be discovered was the green praseo (*trans*) chloride, prepared by Gibbs and Genth (1857a) by the decomposition of $[Co(NH_3)_5H_2O]_2(SO_4)_3$. In 1871 F. Rose prepared the same compound by air-oxidation of an ammoniacal cobalt(II) chloride solution and assigned it the formula $Co(NH_3)_4Cl_3 \cdot H_2O$. In 1877 and 1882 Vortmann prepared the praseo nitrate, chloride and mercury(II) chloride double salt. The most extensive investigations of the praseo series were carried out by Jørgensen (1897) and by Werner and one of his earliest *Doktoranden*, Arnold Klein. Jørgensen prepared the chloride by dissolving in sulfuric acid either $[Co(NH_3)_4(H_2O)Cl]Cl_2$ or $[Co(NH_3)_4(H_2O)_2]_2(SO_4)_3$ and adding concentrated hydrochloric acid. Werner and Klein (1897b) prepared the acid sulfate, which is the easiest compound of the series to obtain in the pure state, by the action of concentrated hydrochloric acid on $[Co(NH_3)_4CO_3]Cl$, followed by treatment with sulfuric acid. They also prepared the praseo chloride, bromide, iodide, fluoride, nitrate, nitrite, thiocyanate, dichromate, hexacyanoferrate(III), hexacyanochromate(III), tetrachloroaurate(III), tetrachloroplatinate(II) and hexachloroplatinate(IV).

The blue violeo (*cis*) dichlorotetraammines were first prepared by Werner (1907 a), who treated

$$\left[(NH_3)_4 Co \underset{HO}{\overset{OH}{\diagup\diagdown}} Co(NH_3)_4 \right] Cl_4$$

with hydrochloric acid saturated with hydrogen chloride gas at $-12\ °C$. Werner thus obtained the *cis*-chloride, bromide, iodide, sulfate and dithionate. Treatment of carbonatotetraamminecobalt(III) chloride with concentrated aqueous hydrochloric acid produces only a small yield of violeo chloride, but Werner later described an improved procedure involving the action of absolute ethanol saturated with hydrogen chloride gas at $0\ °C$ on $[Co(NH_3)_4 CO_3]Cl$ (Werner, 1912). The product is always contaminated with the praseo (*trans*) isomer. Pure violeo salts are usually prepared by means of the insoluble *cis*-dithionate. The preparation of isomerically pure *cis*-$[Co(NH_3)_4 Cl_2]Cl$ in quantitative yield by the action of concentrated hydrochloric acid on *cis* (flavo)-$[Co(NH_3)_4 (NO_2)_2]NO_3$ at $-10\ °C$ has also been reported (Duval, 1926).

Werner's assignment of a *cis* configuration for the violeo dichlorotetraammines was based on their preparations from the ring-containing – and therefore *cis* – compounds

$$\left[(NH_3)_4 Co \underset{O}{\overset{O}{\diagup\diagdown}} C{=}O \right] X$$

$$\left[(NH_3)_4 Co \underset{HO}{\overset{OH}{\diagup\diagdown}} Co(NH_3)_4 \right] X_4$$

and the reasoning is similar to that for the *cis*-$[Co(en)_2 Cl_2]X$ compounds (Werner, 1912). Werner's resolution of the more stable *cis*-$[Co(en)_2 Cl_2]X$ salts, to which the violeo tetraammines are strictly analogous, supports his configurational assignment. Further configurational proof was furnished by X-ray spectroscopic investigations, according to which complexes that exist in two stereoisomeric forms show different X-ray absorption spectra of the chlorine atoms (Stelling, 1927, 1928). The configurations of the praseo and violeo chlorides are as follows:

Praseo (*trans*: 1,6)   Violeo (*cis*: 1,2)
Gibbs and Genth (1857 a)   Werner (1907 a)

The dichlorotetraammines are unstable in aqueous solution and undergo aquation, which is accompanied by changes in color. The blue solutions of the violeo salts rapidly become violet, while the originally green solutions of the praseo salts become violet gradually or more rapidly when warmed. The aquation can be followed by measurements of conductivity, of flocculating power and by spectrophotometry, and the kinetics* for the *trans* compounds have been studied. In general, the *trans* dichloro salts are converted into *trans*-$[Co(NH_3)_4(H_2O)_2]X_3$ salts, with intermediate formation of *trans*-$[Co(NH_3)_4(H_2O)Cl]X_2$ salts. Only the *trans*-acid sulfate does not undergo aquation. In the presence of concentrated acids the violeo salts are converted into the praseo salts. The two series of salts have been separated chromatographically* on alumina. [See FUNDA-MENTALS: *Separation in Chemistry*, in this series.] Werner finally was able to explain the formation of *trans* compounds from *cis* compounds by his theory of rearrangements (Werner, 1912), and consequently the formation of praseo salts by the action of hydrochloric acid on *cis*-$[Co(NH_3)_4(NO_2)_2]X$ or $[Co(NH_3)_4CO_3]X$ mentioned above (p. 116) was no longer regarded as anomalous.

(c) *cis*- and *trans*-Dinitrotetraammines (flavo and croceo salts), $[Co(NH_3)_4(NO_2)_2]X$. The yellow dinitrotetraamminecobalt(III) salts represent the second longest known case of geometric isomerism among coordination compounds. Consequently, they have been extensively investigated, and numerous preparative procedures have appeared in the literature. The orange-yellow *trans* compounds were first prepared by air-oxidation of a solution of cobalt(II) chloride containing ammonium chloride, ammonia and sodium nitrite by Wolcott Gibbs (1875), who named them croceo salts, while the brownish yellow *cis* compounds were first prepared by treating carbonatotetraamminecobalt(III) salts with sodium nitrite by Jørgensen (1894a), who named them flavo salts. The two series correspond completely to the analogous $[Co(en)_2(NO_2)_2]X$ salts.

Both series of compounds were found to possess similar properties. The two nitro groups are 'masked' within the coordination sphere; they are not removed during metathetical reactions, and they resist the action of dilute acids. Further evidence for the strong bonding of the NO_2 groups to the cobalt atom is the existence of the complex in solution as a monopositive ion (and hence the presence of two ions in solution), as demonstrated by cryoscopic measurements by Werner and Herty (1901) and others, by measurements of conductivity by Werner and Miolati (1893) and others, by combined studies of transport numbers and conductivities and by measurements of flocculating power on colloidal solutions.

Concentrated hydrochloric acid converts the croceo salts to *trans*-$[Co(NH_3)_4(NO_2)Cl]Cl$, whereas with the flavo salts both NO_2 groups are replaced, resulting in either *cis*(violeo)-$[Co(NH_3)_4Cl_2]Cl$ (Duval, 1926) or *trans*(praseo)-$[Co(NH_3)_4Cl_2]Cl$. Because of this difference in the reactivity of the NO_2 groups in the two series, Jørgensen regarded them as structural isomers, considering croceo salts as nitro ($Co-NO_2$) compounds and flavo salts as nitrito ($Co-O-N = O$) compounds:

        Flavo   Croceo

Werner argued that flavo salts could not be nitrito compounds since $[Co(en)_2(ONO)_2]X$ salts had been prepared and found to exhibit entirely different properties, being red and acid-sensitive. He thus considered both flavo and croceo salts to be true nitro compounds, differing only in the orientation of these groups in space, i.e. *cis-trans* stereoisomers:

          Flavo (*cis*)   Croceo (*trans*)

Jørgensen felt that Werner was being inconsistent in agreeing with him that xantho ($[Co(NH_3)_5NO_2]X_2$) and isoxantho ($[Co(NH_3)_5ONO]X_2$) salts are structural isomers (see pp. 94–95) yet insisting that flavo and croceo salts were stereoisomers rather than structural isomers. If the isomerism arose as a consequence of purely geometric considerations, Jørgensen argued that other isomer pairs of type MA_4B_2 should exist. Furthermore, he pointed out that Werner's formulation of the croceo salt as a *trans*-dinitro compound required the two nitro groups to be identical, yet one of these groups is attacked much more readily by hydrochloric acid than the other. Misinterpreted evidence based on salt interconversions led Jørgensen to postulate that flavo and croceo salts belong to the praseo and violeo series, respectively, in direct opposition to Werner's views, which were based on direct conversion of praseo salts to croceo salts using sodium nitrite. Subsequently, Werner's theory of rearrangements explained many such inconsistencies (Werner, 1912). Werner's view of the flavo-croceo isomerism eventually prevailed.

The *cis* salts are generally much more soluble than the *trans* salts. The two series can be distinguished by the action of various reagents, in addition to the concentrated hydrochloric acid mentioned above (Jørgensen, 1894). For example, 50% nitric acid converts the *cis*-nitrate to *cis*-[Co(NH$_3$)$_4$(H$_2$O)$_2$](NO$_3$)$_3$ and the *trans*-nitrate to *trans*-[Co(NH$_3$)$_4$-(H$_2$O)NO$_2$](NO$_3$)$_2$. Furthermore, ammonium oxalate or hexafluorosilicic acid gives a precipitate with the *cis* salts but not with the *trans* salts. In the catalytic decomposition of H$_2$O$_2$, the *trans* salts are more active than the *cis*. The *cis* and *trans* isomers have been separated by ion exchange chromatography. The absorption spectra, both in solution and in the solid state, are very similar for both series. The *cis* or *trans* structures are preserved during metathetical reactions, and it has not yet been found possible to convert one series directly into the other.

Optical isomerism
Even though the discovery of the long-sought violeo salts in 1907 convinced Jørgensen that his own views and those of his mentor Blomstrand could not be correct, Werner's success in preparing two, and only two, isomers of the [Co(NH$_3$)$_4$Cl$_2$]X salts as well as numerous compounds of the types mentioned above was not sufficient to prove conclusively his proposed octahedral configuration. Despite such 'negative' evidence, it could still be argued logically that failure to isolate a third isomer of these compounds did not necessarily prove their nonexistence. A more 'positive' proof was necessary. This proof involved the resolution into optical isomers of certain types of asymmetric coordination compounds containing chelate groups (Kauffman, 1975d, 1975e). The agreement of Werner's empirical results with the predictions of his octahedral hypothesis can clearly be seen by comparing Column V with Columns I–IV in Table V.

The concept of asymmetry and resultant optical activity has played an important and venerable role in organic chemistry. If we consider modern organic chemistry to begin with Wöhler's synthesis of urea in 1828 (Kauffman, 1978b, 1979a, b), then Biot's discovery of optical activity in 1812 antedates the very genesis of this field. Furthermore, Le Bel and Van't Hoff's concept of the tetrahedral carbon atom, which constitutes the foundation of organic stereochemistry, was proposed in 1874 largely to explain the optical isomerism investigated by Pasteur and others (Kauffman, 1975g, 1977b). Werner was trained primarily as an organic chemist, and the lecture notes for his course in stereochemistry as well as his *Lehrbuch der Stereochemie*, published in 1904, bear witness to his familiarity with asymmetry and optical activity.

Unfortunately, we do not know exactly when Werner first realized that one of the geometric consequences of his octahedral model was molecular asymmetry for certain types of complexes containing chelate ligands or when he first recognized that a resolution of such compounds would provide an elegant and definitive proof of his stereochemical view that

cobalt(III) possesses an octahedral configuration. Contrary to common
belief, no mention of this topic appears in his first paper on the coordination theory.

According to Victor L. King (1886–1958) (Plate 17), Werner's
American *Doktorand* who successfully solved the problem, Werner and a
series of his students had been attempting to resolve coordination compounds for 'over a period of some nine years', which would date Werner's
first experiments from about 1902. Our first direct, documented evidence,
however, that Werner was actively engaged in experimental attempts to
resolve coordination compounds is found in a letter of 20 February 1897
to his good friend and former fellow student and collaborator Arturo
Miolati: 'At present we are searching for asymmetrically constructed
cobalt molecules. Will it be successful?' We thus see that King's estimate
of the time expended by Werner on the problem was a conservative one.

In 1899, in a paper dealing with oxalatobis(ethylenediamine)cobalt(III)
salts, Werner considered the possibility of optical isomerism among coordination compounds for the first time in print:

On the basis of the octahedral formula, spatial consideration of

the radical $Co\begin{smallmatrix} C_2O_4 \\ \\ en_2 \end{smallmatrix}$ leads to interesting consequences in regard

to the appearance of a new possibility for isomerism. By analogy we must conclude that the most probable bonding of the

residue $\begin{array}{c} O:C\cdot O^- \\ | \\ O:C\cdot O^- \end{array}$ will be the one in the edge position of the octa-

hedron and not the one in the diagonal position, that is:

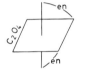 and not

The model resulting from this assumption, however, is, stereochemically speaking, an asymmetric one; i.e. it can be construed in two spatial arrangements which behave as image and mirror image and which cannot be made to coincide.

The case of isomerism developed here is not comparable to the usual asymmetry in organic molecules which, as is well known, is stipulated by so-called optical isomerism, inasmuch as the groupings (2 ethylenediamines) which are here arranged right or left are identical. The above isomerism would rather be comparable to that of organic double ring systems; e.g. of the following type:

which can likewise be construed in two nonsuperimposable models acting as image and mirror image even though no asymmetric carbon atom is present. Among carbon compounds too, this type of asymmetric isomerism has until now not been observed. Thus, for the oxalatodiethylenediaminecobalt salts and similar compounds, we may predict a new type of isomerism which belongs to the class of asymmetry isomerism, of which until now the usual carbon asymmetry and the molecular asymmetry of the inositols are known (Werner and Vilmos, 1899).

Eight years later, success had still not been attained, but Werner had not abandoned his efforts to reach his goal. 'I see from your beautiful paper in *Berichte* that you have been more successful in resolving propylenediamine than we have', he wrote on 15 November 1907 to the Russian chemist Lev Aleksandrovich Chugaev, an enthusiastic supporter of the coordination theory. 'Now I wish to ask whether you would permit me to use the active propylenediamine in the investigation of compounds

$$\begin{bmatrix} O_2N \\ \qquad Copn_2 \\ O_2N \end{bmatrix} X$$

of which we have already obtained five inactive series.'

When King arrived in Zürich, he was assigned the task of resolving carbonatobis(ethylenediamine)cobalt(III) bromide. This salt had been the object of at least one previous documented attempt at resolution, for among the thousands of samples in the collection of Werner's complexes preserved at the Anorganisch-Chemisches Institut der Universität Zürich I found one labeled 'Resolution experiment on

$$\begin{bmatrix} \qquad CO_3 \\ Co \\ \qquad en_2 \end{bmatrix} Br$$

by means of silver-*d*-tartrate, 20/I. 1908, Dubský'. Under the date January 1910 we find as the first entry in King's laboratory notebook the following:

The salt di-ethylenediamine carbonato cobalti bromide is usually represented as follows: $[Coen_2 CO_3] Br$ or

The above salt molecule is not *deckbar* [superimposable] with its *Spiegelbild* [mirror image] and should consist of an equivalent mixture of optically active isomers. The object is to demonstrate the truth or fallacy of this and, if possible, separate the optically active isomers.

(a) The first resolved coordination compound. After a year's unsuccessful attempts to resolve [Co(en)$_2$CO$_3$]Br, King abandoned it and began work on a related compound, *cis*-chloroamminebis(ethylenediamine)-cobalt(III) chloride, [Co(en)$_2$(NH$_3$)Cl]Cl$_2$, a compound that was first discovered by Jørgensen (1890b). As mentioned previously, it is one of those ironies of history that many of the compounds that played crucial roles in the victory of the coordination theory over the rival Blomstrand—Jørgensen chain theory were first prepared by Jørgensen himself.

On p. 16 of King's laboratory notebook we read:

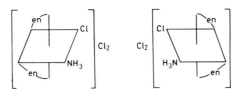

By removing Carbonato Group and placing 2 dissimilar groups NH$_3$ and Cl the *vermutete undeckbarkeit* [expected nonsuperimposability] of the *Spiegelbilder* [mirror images] will be enhanced. By means of the camphor sulfonates perhaps these salts may be separated.

Werner's private assistant Ernst Scholze (PhD, Universität Zürich, 1911) was simultaneously attempting to resolve the corresponding bromo compounds — *cis*-bromoamminebis(ethylenediamine)cobalt(III) salts, [Coen$_2$(NH$_3$)Br]X$_2$, for on p. 21 of King's notebook (undated, but sometime between April 1911 and 12 June 1911) appears the statement: 'With the *Bromoamine Reihe* [bromoammine series] the dextro salt falls right out and no fractionation is necessary. (Werner's Private Lab.)' By 12 June, King had obtained the dextro diastereoisomer of his compound and shortly thereafter the levo diastereoisomers as well:

> I shall never forget the day that the optically active isomers were first attained. In connection with this work, I had been carrying out some 2000 fractional crystallizations and had been studying Madame Curie's work on radium for that purpose. After having made these 2000 separate fractional crystallizations which proved that the opposite ends of the system were precisely alike and that we had to do something more drastic, I proposed increasing the dissimilarity of the diastomers [*sic*] by using brom [*sic*] camphor sulfonic acid as a salt-forming constituent having extremely high optical activity. When this was tried, the isomers in the form of these salts literally fell apart (King, 1942).

King, who was accustomed to being greeted on the streets of Zürich with the inquiry, '*Nun, dreht es schon?*' (Well, does it rotate yet?), continues by recalling how he walked into Werner's office with the long-awaited news. Werner 'leaned back in his chair, smiled, and said not a single word'. The tetrahedron had been forced to relinquish its monopoly on optical isomerism.

All the students knew that something extraordinary must have happened when Werner, who was known for his punctuality, did not appear at his five o'clock lecture. To everyone's astonishment, a young student appeared and announced that the lecture had been canceled. Werner, fearful that the antipodes might racemize overnight, worked late into the night with King, making many derivatives and observing their rotations. But his fears were unfounded, for the enantiomorphs* proved to be remarkably stable.

Werner's immense excitement and pleasure were communicated not only to his students but touched many people, although indirectly. Peter Debye, the late Nobel laureate in chemistry for 1936, then Professor of Physics at the Universität Zürich, recalled:

> One early afternoon when I went from the lake to the Physics Institute after lunch, Werner hailed me from the opposite side of the Rämistrasse. It turned out that he wanted to talk to me about the fact that he had succeeded in making a coordination compound which showed rotation of the plane of polarization. I was very much interested indeed but did not quite understand why he talked to me, since we had had no scientific discussions at all before that time.

During that spring of 1911, many other persons must have been startled by the atypical and unusual behavior of Alfred Werner elatedly accosting casual acquaintances on the street to relate to them the story of his greatest experimental triumph, a work which John Read, a former *Doktorand* of Werner's, called a 'stereochemical achievement of the first order'.

As might be expected, Werner lost no time in submitting King and Scholze's results to the *Berichte der Deutschen Chemischen Gesellschaft*. In his classic paper, which was received on 24 June 1911, little more than a week after King's success, Werner used stereochemical arguments reminiscent of Van't Hoff's paper of 1874 on the asymmetric carbon atom (Werner, 1911). After citing several consequences of the octahedral hypothesis that are amenable to experimental verification, such as the occurrence of complex ions $[MA_5 B]$ in only one form and the occurrence of complex ions $[MA_4 B_2]$ and $[MA_4 BC]$ in two isomeric series, Werner discussed a much more decisive proof – the existence of the optical isomers required by the octahedral model. For the complex [MA₃BCD] in which the three As or B, C and D occupy the three corners of an octahedral face (*cis*; 1, 2, 3), the mirror images are not superimposable:

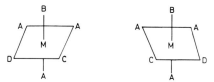

The same is true of the complex [MABC₂D₂] if the groups are arranged in the following manner:

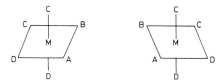

The number of different groups that are required to produce molecular asymmetry is reduced considerably when the coordinated groups are united by bridges, i.e. when monodentate groups are replaced by bidentate groups. (Until 1965, no one succeeded in resolving a complex containing monodentate groups only.) Werner demonstrated that if the groups CC and DD in the last structure are each replaced by a molecule of ethylenediamine, the molecular asymmetry is maintained in the resulting complex *cis*-[MABen₂]. Consequently, compounds with this structure should contain an asymmetric central atom and should therefore be resolvable into optically active antipodes:

For his resolutions Werner had chosen the most widely used method, the racemic modification method developed by Pasteur, in which a solution of the racemic mixture is treated with a resolving agent, that is, an electrolyte containing optically active ions of charge opposite to that of the racemic ions. The two resulting combinations of the ions of the resolving agent and the oppositely charged ions of the racemic mixture are not enantiomorphs, but diastereoisomers, which differ in solubility and other properties and can be separated by fractional crystallization or precipita-

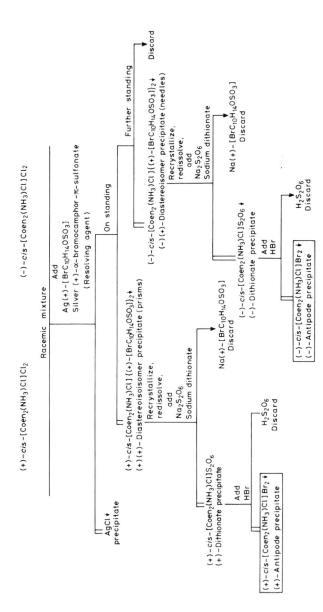

Figure 9.   The first resolution of a coordination compound (simplified scheme).

tion. The optically active ion of the resolving agent is then removed by another reaction, leaving the separated enantiomers of the original racemic mixture pure and free of optically active contaminants.

The method of diastereoisomer formation, although general in principle, often failed in practice, largely because naturally occurring optically active acids and bases are weak and their salts are not very stable in solution. In 1893, the same year in which the coordination theory was published, F. Stanley Kipping and William Jackson Pope had synthesized the strong acids, (+)-camphorsulfonic acid and (+)-bromocamphorsulfonic acid, from naturally occurring (+)-camphor and thus had provided the stereochemist with a series of versatile resolving agents. It was the silver salt of the latter compound that brought Werner his widely acclaimed success. The process is illustrated schematically for King's salts in Fig. 9.

The salts of the bromoammine series are easier to resolve because their diastereoisomeric (+)-bromocamphorsulfonates exhibit a great difference in solubility (Kauffman, 1974a, 1976c). For both series, the (+)-bromo-camphorsulfonate of the (+)-cation, which is less soluble than that of the (−)-cation, crystallizes out first. The diastereoisomers were converted into the bromides either through intermediate conversion to the dithionate or directly by grinding with concentrated hydrobromic acid. For the bromides, Werner obtained the specific rotations $[\alpha]_c = \pm 43°$ (chloro-ammine) and $\pm 46.25°$ (bromoammine).

Contrary to Werner's expectations, the compounds proved to be remarkably resistant to racemization in solution, both on prolonged standing at room temperature and even on heating to incipient boiling. He was able to interconvert various salts within each series without loss of activity. Even more remarkably, he succeeded in replacing the coordinated bromine atom in the bromoammine bromide with water: cis-[Coen$_2$(NH$_3$)Br]Br$_2$ + 3AgNO$_3$ + H$_2$O → cis-[Coen$_2$(NH$_3$)H$_2$O]-(NO$_3$)$_3$ + 3AgBr↓, and he found that the resulting aquaamminebis-(ethylenediamine) salts were optically active.

Only the results of the first 39 pages (up to June 1911) of King's work out of a notebook of 115 pages (up to 1912) were incorporated into Werner's publication, but most of King's work is included in his dissertation. King reported his results up to 1 August 1911 to Werner for use in Werner's lecture 'Über optisch-aktive Kobaltverbindungen', delivered before the Schweizerische Naturforschende Gesellschaft at Solothurn, Switzerland that same day, but the lecture was not published. Thus the majority of King's work, including the resolution by means of the (−)-form of the resolving agent, the preparation of nine forms of the diastereoisomers, and syntheses of the optically active chlorides, nitrates and sulfates, has never appeared in the literature.

According to Werner, the investigation proved that 'metal atoms can act as central atoms of stable, asymmetrically constructed molecules [and] that pure molecular compounds can also occur as stable mirror image isomers, whereby the difference between valence compounds and molecular compounds, which is still frequently maintained, disappears entirely' (Werner, 1911a). Moreover, it confirmed 'one of the most far-reaching conclusions of the octahedral formula'. King considered the resolution to be 'the last proof for the octahedral formula assumed by A. Werner'. He attributed the optical activity to mirror image isomerism rather than exclusively to the presence of an asymmetric atom: 'Whereas until now only a few carbon compounds with such steric structure are known, a considerable number of such metal–ammines have already been successfully prepared.'

The resolution of optically active coordination compounds, a feat which, in Paul Karrer's words, 'shook chemistry to its innermost foundations', gained for the coordination theory the widespread recognition for which Werner had been striving for so long. Nor was the theory's founder neglected, for two years later, largely in recognition of 'the most brilliant confirmation of [his] stereochemical views', as Israel Lifschitz has described the resolution, Werner was awarded the Nobel Prize in chemistry.

Since Werner was originally an organic chemist with extensive experience in stereochemistry, the question as to why it took so long for him and his students to resolve coordination compounds successfully is an intriguing one to which no definite answer is available, but which is a fruitful subject for speculation (Kauffman, 1968, 1974a, 1975d, 1975e, 1976c).

Concerning Werner's choice of a resolving agent, we should note that the common resolving agents such as (+)-tartaric acid do not work for most complexes, and (+)-α-bromocamphor-π-sulfonic acid, the resolving agent which finally proved to be successful, was not readily available at the time. Since its synthesis is a very long and tedious process, Werner and his students probably did not try it until they had exhausted all other possibilities.

Concerning Werner's choice of the compound to be resolved, since many anionic complexes such as the oxalato compounds racemize rather readily, Werner might have missed these resolutions if he had attempted any. He and his students may have made numerous unfortunate choices of resolving agents and complexes, perhaps sometimes achieving resolutions, but not recognizing them because of rapid racemization.

Improvements in instrumentation that occurred in the first decade of the present century might have played the largest role in Werner's success. Before this time, polarimeters*, color filters and light sources were still in too primitive a state to permit easy observation of the deeply colored

Plate 18. Werner's polarimeter (Franz Schmidt and Haensch No. 8142)
[Photograph by Herr Richard Taubenest, Universität Zürich]

solutions with which Werner was forced to work. It was only shortly before the time of Werner's success that the Schmidt and Haensch Model
No. 8142 polarimeter (Plate 18) became commercially available. This
instrument, which brought Werner his long-sought success, was based
upon the specifications of Hans Heinrich Landolt and included several
new features of importance for measuring small rotations, especially the
Lippich *Halbschatten* (half-shadow) device which better defined the uni-
form gray appearance midway between the extinctions of the two halves
of the field and thus increased the precision in determining the zero
points.

Another factor that may have proved crucial was Werner's choice of the
wavelength of the light used for his measurements. Most of the rotations
at that time were measured at the wavelength of the sodium D-line produced by thermal excitation (589 nm), but the intense color of Werner's
solutions would have rendered use of this wavelength difficult. At about
the time of Werner and King's success, Landolt had described the use of
filters to obtain different wavelengths. By employing a pair of colored
solutions (crystal violet and potassium chromate) as a filter for his Nernst
projection lamp, Werner obtained a light of optical mass-center 665.3 nm,
very close to the wavelength of the Fraunhofer C-line (656.3 nm), and all
of his optical rotation values were reported as $[\alpha]_C$ rather than $[\alpha]_D$.

Finally, as anyone who has ever attempted it will attest, resolution of
optical isomers in the absence of exact directions is as much an art as a
science. It is often a time-consuming and tedious process for which trial-

and-error methods and patience may be the only solutions. In the development of such methods, experimental details may be extremely critical, as for example in the first resolution ever performed, the mechanical separation of the hemihedral crystals of sodium ammonium tartrate which Pasteur obtained by spontaneous evaporation of the solution. The two antipodes unite to form a racemate at temperatures above 26 °C so that Pasteur would probably not have made his discovery had he been working in a warm Mediterranean climate rather than the cool Parisian one (Kauffman, 1975d, 1975e, 1975g, 1977b).

Experimental conditions may be just as crucial in the resolution of coordination compounds. When Robert W. Auten attempted to resolve the $cis[Co(en)_2Cl_2]^+$ ion according to Werner's procedure, he obtained no diastereoisomer precipitate at all, even after repeating Werner's directions to the letter several times. Reasoning that Werner had worked in Zürich where it is not customary to heat the laboratories as much as in the United States, Auten repeated the experiment with the solutions cooled to 16 °C (Bailar and Auten, 1934). The solubility curve for this particular diastereoisomer must be a very steep one, for at 16 °C he got an almost quantitative yield of diastereoisomer in agreement with Werner's results, whereas at 22 °C he had obtained no precipitate at all.

Although our speculations have not yielded a definite answer to our original question, perhaps they have persuaded us to rephrase the question. Rather than asking what took Werner so long, perhaps we should marvel that he succeeded at all in the time that he did. If we consider the possible variety of methods, complexes, resolving agents, instruments and wavelengths as well as the highly specific experimental conditions sometimes required for success in such ventures, it is not unlikely that even for someone with Werner's chemical intuition and experimental skill, many years might be required to solve the problem. In the final analysis, Werner's conclusive proof of the octahedral configuration for cobalt(III) by the resolution of coordination compounds was made possible by his unshakable faith in his own ideas and his persistent and untiring efforts to prove them, even in the face of what might have seemed to others to be unsurmountable experimental difficulties.

(b) A completely inorganic optically active complex. 'Whenever Werner opened up a new field, he expanded it with unbelievable speed'. This statement of Paul Karrer's was amply confirmed by Werner's investigations of optically active complexes, for once Werner had found the key to the resolution of complexes, a large number of articles describing additional resolutions appeared from his institute with great rapidity (Kauffman, 1968, 1974b). Within eight years, he and his students had resolved more than forty series of cationic and anionic complexes, not only of

cobalt but of other hexacoordinate transition metals as well. In this way he succeeded in proving the octahedral configuration for iron(II), chromium(III), rhodium(III), iridium(III) and platinum(IV). In 1913 he even repaid his debt to Pasteur by using optically active inorganic octahedral complexes, which had been resolved by means of organic substances, to resolve in turn dimethylsuccinic acid, an organic tetrahedral compound. In the same year, he also proved that polynuclear as well as mononuclear complexes could be resolved and thus demonstrated the theoretically predicted analogy between compounds containing two asymmetric carbon atoms and polynuclear complexes with two metal atoms, another striking confirmation of his octahedral hypothesis. In complete analogy with tartaric acid, which, in addition to the racemic (+)(−)-form, also exists in (+)- and (−)-enantiomers and in an internally compensated, nonresolvable (meso) form (see pp. 47–49), Werner was able to demonstrate experimentally for the binuclear complex

$$\left[en_2Co \underset{NO_2}{\overset{NH_2}{\diagdown\diagup}} Coen_2 \right] X_4$$

the existence of a racemic (+)(−)-form, (+)- and (−)-enantiomers, and an internally compensated, nonresolvable (meso) form (see p. 48).

Although the compounds that Werner had resolved up to 1914 represented a remarkable variety of compound types, they all possessed one common characteristic — they all contained carbon. Because of the then prevalent view that optical activity was almost always connected with carbon atoms, a number of Werner's contemporaries could argue that the optical activity of all these compounds was somehow due to the ethylenediamine or bipyridyl molecules or to the oxalate ions contained in them, even though these symmetrical ligands are themselves optically inactive. By such devious reasoning, they were able to cast doubt on the validity of the octahedral configuration. In 1914 Werner was able to silence even the most skeptical of his opponents and to vindicate unequivocally his octahedral concept. In that year he resolved a completely carbon-free coordination compound of the $M(\overline{AA})_3$ type, viz. tris[tetraammine-μ-dihydroxocobalt(III)]cobalt(III) bromide:

$$\left[Co \left\{ \underset{HO}{\overset{HO}{\diagdown\diagup}} Co(NH_3)_4 \right\}_3 \right] Br_6$$

a compound that, ironically enough, had been first discovered sixteen years previously by Jørgensen (1898).

The lustrous brown tetranuclear tris[tetraammine-μ-dihydroxo-cobalt(III)]cobalt(III) salts were first discovered and named *Anydrobasische Tetrammindiaquodiamminkobaltsalze* by Jørgensen. Werner *et al.* (1907 c) showed that they possess the constitution shown above. Since they are thus structurally similar to $[M(en)_3]X_n$ salts with

$$\left[\begin{array}{c} HO \\ \\ HO \end{array} \hspace{-0.5em} \begin{array}{c} \\ \diagdown \\ \diagup \end{array} \hspace{-0.5em} Co(NH_3)_4 \right]^+$$

ions in place of ethylenediamine molecules, they should be capable of existing in nonsuperimposable mirror image forms.

Werner succeeded in resolving the bromides by diastereoisomer formation with silver (+)-bromocamphorsulfonate, but the process was extremely tedious because of the small differences in solubility between the diastereoisomers. Optical measurements were hindered by rapid racemization and the deep color of the solutions. Rotations were measured in 50% aqueous acetone solutions in order to minimize racemization. The highest values found were $[\alpha]_{560} = -4500°$, $[M]_{560} = -47, 475°$ for the pure (−)-bromide and $[\alpha]_{560} = +1050°$, $[M]_{560} = +11, 109°$ for the (+)-bromide. Although the structure of these compounds has not yet been determined by X-ray diffraction, the structure of the corresponding ethylenediamine cation has been so determined and found to agree with Werner's formulation. To the present day, with the sole exceptions of R. D. Gillard and F. L. Wimmer's resolution (1978) of $[Pt(S_5)_3]^{2-}$ and Satoru Shimba, Shuhei Fujinami and Maraji Shibata's resolutions (1979) of *cis-cis-cis*-$[Co(NH_3)_2(H_2O)_2(CN)_2]^+$ and *cis-cis-cis*-$[Co(NH_3)_2(H_2O)_2$ $(NO_2)_2]^+$, F. G. Mann's resolution of (1933) of Na *cis*-$[Rh(H_2O)_2[SO_2 (NH_2)_2]_2]$, Werner's resolution of

$$\left[Co \left\{ \begin{array}{c} HO \\ \\ HO \end{array} \hspace{-0.5em} \begin{array}{c} \\ \diagdown \\ \diagup \end{array} \hspace{-0.5em} Co(NH_3)_4 \right\}_3 \right] X_6$$

remains the only example of the resolution of a completely carbon-free coordination compound and marks his crowning achievement in coordination chemistry.

In Werner's own words, the investigation proved that 'carbon-free inorganic compounds can also exist as mirror image isomers' and that therefore 'the difference still existing between carbon compounds and purely inorganic compounds disappears' (Werner, 1914). At last he had confirmed his long-held view of the unity of all chemistry. The structural theory of organic chemistry was only a special case of the coordination theory, in which the carbon atom happened to have its valence equal to its coordination number. The last brick in the crumbling wall of separation between inorganic and organic chemistry had been razed. The demolition begun 86 years earlier by Friedrich Wöhler with his artificial synthesis of urea from ammonium cyanate in 1828 had been completed by Werner (Kauffman, 1978b, 1979a, 1979b).

By any standards, Werner's achievements are remarkable. At the very beginning of his career, he had destroyed the carbon atom's monopoly on geometric isomerism. In his doctoral dissertation, he had explained the isomerism of oximes as due to the tetrahedral configuration of the nitrogen atom (Werner, 1890). Then, at the peak of his career, he had likewise forced the tetrahedron to relinquish its claim to a monopoly on optical isomerism. He had finally attained one of the major goals of his life's work — the demonstration that stereochemistry is a general phenomenon not limited to carbon compounds and that no fundamental difference exists between organic and inorganic compounds.

During his last years Werner devoted himself almost exclusively to studies of the optically active compounds which had brought him the Nobel Prize and had proved beyond the shadow of a doubt his stereochemical views. Without the impetus of his octahedral hypothesis, who would have thought of looking for optical activity among compounds such as the trioxalato salts, $R_3 \overset{I}{[M}\overset{III}{(C_2O_4)_3}]$, previously regarded merely as double salts $3R_2 \overset{I}{(C_2O_4)} \cdot M_2 \overset{III}{(C_2O_4)_3}$ similar to the alums? Werner's investigations of the optically active coordination compounds of cobalt, chromium, iron, rhodium, iridium and platinum underlie much of the more recent and sophisticated studies of the thermodynamics; kinetics; visible, ultraviolet and infrared spectra; rotatory dispersion; circular dichroism; ligand exchange; racemization; and absolute configuration of these and similar compounds.

The validity of Werner's structural views was later amply confirmed by numerous X-ray diffraction studies (see pp. 151—155). Yet, despite the introduction of more *direct* modern techniques, his classical configurational determinations by simple *indirect* methods still remain today a testament to his intuitive vision, experimental skill and inflexible tenacity.

In the words of former American Chemical Society President Henry Eyring:

The ingenuity and effective logic that enabled chemists to determine complex molecular structures from the number of isomers, the reactivity of the molecule and of its fragments, the freezing point, the empirical formula, the molecular weight, etc., is one of the outstanding triumphs of the human mind (*Chem. Eng. News*, 7 January 1963, p. 5).

Although some of Werner's methods of resolution have been improved, and his specific rotation values for many complexes have been shown to be too low, we must always bear in mind that he was the pioneer who first opened the door to a previously unsuspected field. In his last works he stood on the threshold of an extremely complicated research area — the investigation of optically active coordination compounds containing optically active ligands. Had the powerful, creative trend of his life not been cut short by his untimely death, there is no telling what Alfred Werner might have accomplished in this field.

In the next chapter we shall examine some of the developments in coordination chemistry that have taken place in our own century — discoveries and concepts that would have been unthinkable without Werner's pioneering work that preceded them.

7

Coordination chemistry in the twentieth century

INNER COMPLEXES: LEY AND BRUNI (1904)

Werner's theory (1893) represented a watershed in the history of coordination chemistry. In fact, it was Werner's coordination theory that gave the field its name. With a few exceptions such as Jørgensen and Friend, the majority of chemists accepted Werner's views, and most of the twentieth-century contributions to coordination chemistry have been developments, extensions or confirmations of Werner's theory rather than ideas incompatible with or opposed to it. Ley's concept of inner complex salts is one of the earliest of such post-Werner developments (Kauffman, 1973g, 1973h, 1976b, 1978a).

According to Harvey Diehl, 'the idea of the ring structure in ethylenediamine complexes runs subconsciously through the early papers of Werner without being definitely expressed'. In his very first paper on the coordination theory (1893), Werner discussed the structures of the violet (*cis*) and green (*trans*) forms of $[Co(en)_2 X_2]X$, discovered by his scientific adversary Sophus Mads Jørgensen (1889, 1890b). In his treatment of what is regarded as the classic case of geometric isomerism among coordination compounds, Werner recognized that the ethylenediamine molecule occupied two coordination positions. In 1899 Werner and Vilmos described the compound $[Co(en)_2 C_2 O_4]Cl$, in which, in addition to the ethylenediamine molecule, the oxalate anion was regarded as what Sir Gilbert T. Morgan would later call a chelate group (Werner and Vilmos, 1899). Two years later Werner (1901) prepared the nonelectrolyte bis(2,4-pentanedionato)platinum(II), an inner complex of platinum with the enolate anion of acetylacetone:

Plate 19. Friedrich Heinrich Ley (1872–1938) [Courtesy, Herr Heinrich Ley, Oberstudienrat, Lübbecke/Westfalen, German Federal Republic]

in which he postulated a structure involving chelate rings, and as early as 1887, A. Combes, the discoverer of acetylacetone, had prepared acetylacetonates of aluminum, copper, magnesium, iron and lead.

Although metal chelates were thus known in the late nineteenth century, the first person to recognize clearly the special significance and consequences of the cyclic structure in coordination compounds was undisputably the German chemist Henrich Ley, Professor of Chemistry at the University of Münster (1872–1938) (Plate 19). In 1904, by means of observations of color, transference and distribution experiments and determinations of molecular weight and electrical conductivity, he explained the constitution of copper glycinate and related compounds by applying Werner's concepts of *Hauptvalenzen* (primary valencies) and *Nebenvalenzen* (secondary valencies) (Ley, 1904). He showed that the compound was not an ordinary simple salt or even an ordinary complex salt but rather a special type of metal chelate that he called an inner metal complex salt (*inneres Metallkomplexsalz*):

$$\left[\begin{array}{c} CH_2{-}NH_2 \quad NH_2{-}CH_2 \\ \quad\quad Cu \\ O{=}C{-}O \quad O{-}C{=}O \end{array}\right]$$

bis(Glycinato)copper(II)

in which each bidentate ligand is bonded to the central metal ion by both a primary valency and a secondary valency, forming a cyclic structure. Ley gave no explanation for the term 'inner', but David P. Mellor assumed that it referred to Werner's inner (nonionizable) coordination sphere as opposed to the outer (ionizable) coordination sphere. Thus an inner complex salt is a coordination compound with all ligands in the inner sphere, i.e. it is a nonelectrolyte, and the term 'salt' is actually a misnomer. A. Liebhafsky interpreted Ley's term as follows: '*inner* connotes ring formation; *complex*, the presence of a second (coordinating) valence; and *salt*, the presence of a primary bond involving a negative group (in this case, $NH_2 CH_2 COO^-$) derived from an acid'.

Ley apparently intended the term inner complex to be restricted to nonelectrolytes. This restriction has been accepted by Diehl and many others but questioned by Liebhafsky and B. O. West. Neutral species are formed whenever the coordination number and charge of the metal ion equal the sum of the donor groups and charges, respectively, of the combining chelate groups. The commonest inner complexes are formed from bidentate ligands with one replaceable hydrogen atom and metals, the coordination numbers of which are twice their ionic charge. Functional groups coordinating to the metal through primary valencies (with displacement of hydrogen) include $-COOH$, $-OH$, $-SO_3H$, $=NOH$, $-NH_2$ and $-NRH$. Those coordinating through secondary valencies (without displacement of hydrogen) include $-NH_2$, $-NH$, $-\overset{|}{N}-$, $=\overset{|}{N}-$, $=NOH$, $-OH$, $=CO$ and $-S-$.

Inner complexes are sometimes divided into two classes – first order and second order – a classification that A. A. Grinberg attributed to Ley himself. Those of the first order are nonelectrolytes, whereas those of second order are electrolytes, e.g.

$$\left[Ti\left(\begin{array}{c} O{=}C{-}R \\ \quad\quad CH \\ O{-}C{-}R \end{array}\right)_3\right]^+$$

Those of the first order have received considerable attention because of their unusual general properties such as great stability and slight degree of dissociation into free metal ions, insolubility in water and solubility in nonpolar organic solvents and anomalous colors differing greatly from the colors of ordinary metal salts. Some of these compounds are so stable that they may be volatilized without decomposition. For example, in 1894 Combes confirmed the atomic weight of beryllium by determining the vapor density of its acetylacetonate. More recently, in one of the most exciting developments of modern analytical chemistry, gas chromatography of volatile β-diketonate inner complexes has been used to separate metals such as beryllium, aluminum and chromium, rare earths and even optical isomers.

Despite the venerable ancestry of metal chelates as dyes and pigments, theories of their formation have been relatively recent. In 1893 Carl Liebermann stated that the formation of mordant dyes was dependent on the production of a 'cycloid' in which the metal or metal complex was included. Chugaev and Werner later criticized Liebermann's theory as being too limited. Werner (1908, 1909) specifically made use of Ley's concepts in his theory of mordant dyes (*Beizenfarbstoffe*). He concluded that the formation of colored lakes (*Farblacke*) depends upon the production of inner complexes and that dyes capable of combining with mordants possess both a salt-forming group (*Hauptvalenz*) and a group that can form a coordinate linkage (*Nebenvalenz*) with a metal atom. Since then, Ley's ideas have been extensively used in the development of new dyes and pigments.

The concept of inner complexes is sometimes associated with the name of the Italian chemist Giuseppe Bruni (1873–1946), who, with C. Fornara, proposed a constitutional formula for copper glycinate that is analogous to Ley's formula. Bruni and Fornara (1904) prepared and examined the copper(II) and nickel(II) salts of glycine, α-alanine, leucine, α-aminoisobutyric acid, aspartic acid and *o-*, *m-* and *p-*aminobenzoic acids. They concluded that the copper salts of aliphatic amino acids differ from most other copper salts in possessing their own blue-violet color that is not changed by addition of ammonia. Since the solutions give few of the reactions characteristic of Cu^{2+}, they assumed that this ion is present only in extremely low concentrations, and they proposed as alternative structures for the glycinate:

Plate 20. Lev Aleksandrovich Chugaev (1873–1922) [Courtesy, the late Academician Il'ya I. Chernyaev, Director, N. S. Kurnakov Institute of General and Inorganic Chemistry of the Academy of Sciences of the USSR, Moscow, USSR]

CYCLIC BONDING AND STABILITY: CHUGAEV'S RULE OF RINGS (1906)

Closely related to Ley's concept of inner complexes is Chugaev's so-called rule of rings. In order for a molecule with two potentially coordinating groups, such as a diamine, $NH_2(CH_2)_n NH_2$, or an amino acid, $NH_2(CH_2)_n COOH$, to function as a chelate group, it must be geometrically possible to form a ring of low strain. From organic chemistry it is known that 5- and 6-membered rings are the most stable, whereas 4-membered rings are less stable and 3-membered rings are quite unstable (Baeyer's strain theory). That these relationships are completely applicable to complex compounds was shown by the Russian chemist Lev Aleksandrovich Chugaev, Professor of Chemistry at St Petersburg University (1873–1922) (Plate 20), as early as 1906 (Chugaev's rule of rings). This rule served as a point of departure for stereochemical research not only by Chugaev but also by many other workers (Kauffman, 1963, 1973j, 1978a). The widely used diagnostic method for determining the configuration of coordination compounds of Pt(II) by reaction with oxalic acid, which was subsequently developed by Chugaev's pupil A. A. Grinberg, provides an excellent example of the stability of a 5-membered ring:

Chugaev clearly showed the stability of pentatomic and hexatomic ring systems by comparing the ease of formation of complexes of Co(III), Ni(II) and Pt(II) with various diamines (Chugaev, 1906). Dimethylenediamine and trimethylenediamine readily form complexes in contrast to higher polymethylenediamines. That 5-membered saturated ring systems are favored over 6-membered systems was shown by the lesser stability of trimethylenediamine compounds compared to the analogous propylenediamine compounds. Also, disulfides and diselenides, R(S or Se)-$(CH_2)_n$(S or Se)R', form complexes with Ni(II) and Cu(II) only when $n = 2$. The tendency toward the formation of 5- and 6-membered rings and the greater stability of the former was further shown by Chugaev and Serbin's study in 1910 of the formation of inner complexes of Cr(III) with various amino acids as a function of the separation of the carboxyl and amino groups. The α-acids formed inner complexes readily, whereas the β-acids did so with some difficulty; inner complexes could not be obtained at all with γ-, δ- and ε-amino acids.

By assuming that ring compounds are more stable than one-chain compounds and that 5- and 6-membered rings are the most stable, Chugaev was able to explain the stabilities of numerous coordination compounds. Thus ammonia, monoamides, hydrazines and γ-amino acids form relatively unstable compounds, whereas ethylenediamines and propylenediamines, amidines and amidoximes, 2,2'-bipyridines and phenanthrolines, α- and β-amino acids, nitrosoguanidines and aminoguanidines and other compounds containing coordinating groups in the 1,2 or 1,3 positions form stable complexes. For the same reasons, compounds containing no nitrogen such as the salts of oxalic, α-oxy aliphatic, carbonic and sulfuric acids are more stable than salts of open-chain monovalent acids.

As early as 1885 M. Il'inskiĭ and G. von Knorre had employed α-nitroso-β-naphthol for the determination of cobalt, but the real impetus for the application of inner complexes to analytical chemistry was provided by Ley's paper, which directed attention to a fruitful area of borderline inorganic-organic chemistry, both pure and applied (Ley, 1904). In 1905 Chugaev made his best known discovery, the reaction of nickel(II) ion with dimethylglyoxime, the first organic spot test reagent used to detect a metal ion. The scarlet precipitate familiar to almost every student of qualitative analysis was only one of many inner complexes investigated by Chugaev and his students that were formed by coordination of metals such as iron, cobalt, nickel, palladium and platinum with α-dioximes. Because these so-called dioximines may be extracted into organic solvents, they are useful in analytical separations. Their intense colors make them extremely sensitive reagents for both qualitative detections and quantitative colorimetric determinations of metal ions, and their insolubility forms the basis for a number of gravimetric methods.

Although the dioximines of cobalt, which were synthesized in great numbers by Chugaev, have no analytical significance, they were of great theoretical importance in the development of coordination chemistry. The Blomstrand–Jørgensen chain theory excluded the possibility of non-electrolytic complexes, whereas the Werner coordination theory predicted the existence of such compounds. Since the number of such stable non-electrolytic complexes was extremely small at the beginning of the twentieth century, the cobalt dioximines and the measurement of their conductances played a crucial role in the triumph of the coordination theory, which Chugaev firmly supported.

The studies on dioximines were intimately connected with Chugaev's rule of rings, which runs through his research on complexes like a unifying thread. A single oxime group shows little tendency toward coordination, but when it can form part of a chelate ring, as in a dioxime, the nitrogen atom becomes a good electron pair donor. Symmetrical dioximes can exist in three isomeric forms:

anti   syn   amphi

Of the three isomers of dimethylglyoxime, Chugaev found that only the *anti*-isomer forms the characteristic scarlet precipitate with Ni(II). He thus was able to use complex formation as a method for determining the isomeric configuration of α-dioximes. The scarlet precipitate was first believed by Chugaev to contain an improbable 7-membered ring (I), but this idea was later revised in favor of a 6-membered ring (II). Paul Pfeiffer's formulation of a 5-membered ring (III) has been supported by X-ray data which showed the presence of multiple rings involving extremely strong hydrogen bonding (IV).

I   II   III

IV

SEVERAL ALTERNATIVE THEORIES OF COORDINATION COMPOUNDS: FRIEND (1908), BRIGGS (1908), POVARNIN (1915) and PFEIFFER (1920)

Werner's stereochemical concepts were amply confirmed by experimental data, much of it provided by him and his students over a period of a quarter-century. However, the widespread dissatisfaction with his concepts of *Hauptvalenz* and *Nebenvalenz* acted as a strong deterrent to his entire theory for a number of years. A clear differentiation between the two types of valence was not always possible, and many chemists considered Werner's ideas of primary and secondary valence bonds vague and unfounded.

Werner invoked his principal valence–auxiliary valence dichotomy in order to explain how apparently saturated metal atoms in salts can combine with additional atoms or molecules to form complex compounds. The two types of valence differed in their origin only, and in the final complex Werner no longer distinguished between them. He was never satisfied with his artificial valence dichotomy, and he sometimes wrote formulae without distinguishing the two types of valence. In 1896 he wrote that principal valence expresses only stoichiometry and gives no information about the number of bonds. In 1907, in an address to the British Association for the Advancement of Science at Leicester, he stated:

The difference between the two kinds of valencies is retained because it seems necessary in view of the present transitional state of the theory to construct sharply defined partial conceptions which can afterwards serve as the foundation stones for a more comprehensive concept of valency (Werner, 1907b).

These more comprehensive concepts have been developed in the electronic theories of valence proposed by Walther Kossel (1888–1956), Gilbert Newton Lewis (1875–1946), Irving Langmuir (1881–1957), Nevil Vincent Sidgwick (1873–1952), Kasimir Fajans (1887–1975), Linus Pauling (b. 1901) and others. The models developed in terms of the electronic theory were so successful in resolving the confusion concerning *Hauptvalenz* and *Nebenvalenz* that almost general acceptance of Werner's theory soon followed the work of Lewis and his contemporaries.

The earliest, most extensive and most protracted attack upon Werner's coordination theory was made, as we have already seen in Chapter 6, by Jørgensen, who was quick to respond to Werner. Werner's coordination theory (dated December 1892) appeared in 1893 (Werner, 1893), and Jørgensen's first critique (dated 13 August 1893) appeared early in 1894 (Jørgensen, 1894). However, it was only one of a number of criticisms of the coordination theory to appear in the literature, some as late as the 1920s. As typical examples, we shall mention here those of Friend, Briggs,

Plate 21.   John Albert Newton Friend (1881–1966) [Courtesy, the late Dr J. A. N. Friend, Birmingham, England]

Povarnin and Pfeiffer. Werner, in his important publications on valence theory, did not respond to any of their criticisms or to those of others.

In 1908 the English chemist John Albert Newton Friend (1881–1966) (Plate 21) criticized Werner's theory and offered an alternative explanation of complex compounds (Kauffman, 1972d, 1978a). Friend noted that the classical theory of valency could not explain why the valency of most elements apparently varies, why electropositive elements readily combine with electronegative ones to form the most stable compounds, why electropositive and electronegative elements combine both with themselves and with other elements of the same sign to form fairly stable molecules to form highly stable complexes (Friend, 1908a). In order to bonding was understood) and why molecules can combine with other molecules to form highly stable complexes (Friend, 1908). In order to resolve these difficulties Friend distinguished three kinds of valency: (1) free positive, (2) free negative and (3) residual or latent valency. His third type referred to positive and negative valencies, which differ from the free valencies in that they can only be called out in pairs of equal and opposite sign. Friend's free valencies correspond to Werner's *Hauptvalenzen* and his latent valencies to Werner's *Nebenvalenzen*, with certain differences in the latter case.

Friend (1908b) specifically criticized Werner's theory and emphasized the differences between his own latent valencies and Werner's *Nebenvalenzen*, viz. groups attached by latent valencies are dissociable whereas those attached by *Nebenvalenzen* are not, and *Nebenvalenzen* differ from *Hauptvalenzen* in energy content whereas latent and free valencies are virtually identical. Friend applied his new theory of valency to ammonium salts, metal–ammines and halide 'double salts', and he proposed that for hexacoordinate central atoms a 'hexatomic shell' forms around the metal but that all the elements or groups are joined *together* by latent valencies and not necessarily to the metal itself.

Eight years later Friend applied his cyclic theory of complexes in detail to the structures of the chlorides of hexaamminecobalt(III), chloropenta-amminecobalt(III) and dichlorotetraamminecobalt(III), and he emphasized four basic differences between his theory and Werner's (Friend, 1916). More than half of this paper was devoted to structural formulae intended to account for the supposed isomerism of potassium ferrocyanide and of potassium ferricyanide, isomerisms that are now known not to exist.

Later that year Eustace Ebenezer Turner criticized Friend's theory as being 'somewhat obscure' on several points. Friend apparently did not respond to Turner's criticism, but in 1921 in a paper entitled 'Electrochemical Conceptions of Valency' Friend attempted to show how his purely chemical theory of valency originally proposed in 1908 was supported by and in harmony with the physical interpretations of the relatively new electronic theory of valency (Friend, 1921). He again suggested 'the necessity of remodelling Werner's theory' and offered additional cyclic structural formulae for various coordination compounds. Friend admitted that his 'shell theory closely resembles spatially the scheme suggested by Werner, in that the groups are arranged at the corners of an octahedron'. According to Friend, it differed, however, in that: (1) it is electrochemical and postulates the existence of only such valencies as are compatible with electronic conceptions; (2) the groups around the central metal atom are joined by latent or nonionized valencies and not necessarily to the central atom itself; (3) it explains many simple and complex inorganic compounds not explicable by Werner's theory; and (4) it shows that the activity of optically active cobalt complexes is due not to an asymmetric arrangement around the cobalt atom, as Werner suggested, but to the presence of an asymmetric nitrogen atom. Friend gave no details on this last point; he promised to 'deal more fully' with it shortly but apparently never did.

Samuel Henry Clifford Briggs (1880–1935), in a paper entitled 'The Constitution of Co-ordinated Compounds', published just after Friend's two papers of 1908, criticized Werner's formulae because 'they tell us so little of the way in which the affinities of the separate atoms are combined, and are consequently far inferior in utility to the valency formulae which have been such an important feature in the development of organic chemistry' (Briggs, 1908). Briggs devised formulae that he claimed fulfilled the conditions required by experimental data and that indicated the manner in which the affinities of the atoms are disposed in the molecule. He did not claim to be proposing a new theory of valency; like Werner, he believed that 'a totally comprehensive and satisfactory theory of valency will not be possible until we have a much more complete knowledge of the constitution of molecular compounds, and also of the nature of chemical affinity, than we possess today' (Briggs, 1908).

Briggs based his formulae on the theory of 'duplex affinity' — the idea that every element possesses two kinds of valency, positive and negative — which was the foundation of the electrochemical system of Berzelius and which had recently been revived by Richard Abegg (1869–1910) in his concept of normal valencies and contravalencies, the sum of which is eight. He represented the saturation of valencies by arrows pointing from positive to negative and unsaturated affinity by a dotted line with an arrow head, pointing away from the atom if the affinity is positive and towards the atom when the affinity is negative. A few examples of Briggs' formulae for platinum(IV)–ammines should suffice:

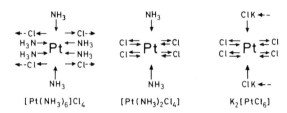

According to Briggs, his formulae provided a simple explanation of the changes in valency and in the charge on the complex, explained the extent of dissociation of complexes in solution and explained all the known cases of isomerism among compounds with an anionic complex.

Briggs later reinterpreted his formulae in terms of the electronic theory of valency, distinguished between what he called primary and secondary affinity, concepts similar to Werner's *Hauptvalenz* and *Nebenvalenz*, and characterized three different types of chemical combination (Briggs, 1917). Like Friend, Briggs also used his formulae to explain the so-called isomeric forms of potassium ferricyanide (α, red; β, olive-green) discovered by Locke and Edwards and even adduced experimental evidence for a similar isomerism among the ferrocyanides. The idea that the alkali metal ferricyanides and ferrocyanides occur in isomeric forms, however, was later disproved by Briggs himself and others.

In 1915, in a series of four papers (the first in two parts) entitled 'Valency of the Elements', the Russian chemist and tanning technologist Georgiĭ Georgievich Povarnin (1880–1946) criticized Werner's theory on the grounds that: (1) it necessitates the assumption of two kinds of valences, with a further differentiation of groups into 'ionogens' and 'non-ionogens', thus making three varieties of affinity bonds; (2) it makes a distinction between atomic and molecular compounds; (3) it assumes that while the number of principal valences of an element depends on its position in the periodic table, the number of its auxiliary valences is not related to the table; (4) in many cases it makes it difficult to predict the number of possible isomers; and (5) it explains only the structure of the inner sphere where atoms or groups are linked directly to the central atom but fails to show how this sphere is united with the outer sphere where union with the central atom is through the intermediacy of other atoms (Povarnin, 1915).

To overcome these difficulties Povarnin proposed his theory of polar affinities, in which he assumed that atoms themselves are complex systems of positive and negative affinities in unequal amounts so that an element is positive when the positive charge exceeds the negative, and negative when the situation is vice-versa. He therefore regarded the classical unit of valency as composed of two polar charges not identical in absolute magni-

Plate 22. Paul Pfeiffer (1875–1951) [Courtesy, Fräulein Crescentia R. Roder, Bonn, German Federal Republic]

tude. He postulated that bonds between atoms have the power to oscillate and that atoms tend to form 4- or 6-membered cyclic molecules.

Povarnin considered his system to offer the following advantages: all bonds are of the same type, there being no distinction between primary and secondary valences; the difference between atomic and molecular compounds disappears, all union being atomic; the prediction of the number of isomers is facilitated because all valences are regulated by the position of the element in the periodic system; and the formation of complexes is readily explained by halving the number of total valences and by the presence of free polar affinities. He applied his theory not only to coordination compounds but also to organic nitrogen compounds, and he even correlated the tanning and swelling of pelts with valence and complex formation.

Even so great an admirer of Werner as the German chemist Paul Pfeiffer, Professor of Chemistry at the University of Bonn (1875–1951) (Plate 22), Werner's former student and one-time 'chief of staff' at the University of Zürich and the man who first applied Werner's theory to crystal structures (see pp. 150–151), proposed modifications of the coordination theory. For example, he applied what he called the principle of 'affinity adjustment of the valencies' to overcome certain shortcomings of Werner's theory (Pfeiffer, 1920). He considered the ionizable radicals or atoms in the outer sphere to be combined with the complex radical as a whole and not attached definitely to the central atom or to any of its associated molecules. He also applied this idea to complex organic molecular compounds. In no way, however, should Pfeiffer's modifications be interpreted as attacks on Werner's ideas.

'CRYSTALS AS MOLECULAR COMPOUNDS': PAUL PFEIFFER (1915)

Crystallography and stereochemistry have always been closely related. Louis Pasteur's resolution of sodium ammonium racemate led to the founding of stereochemistry by Le Bel and Van't Hoff in 1874. August Kekulé, in his 1877 *Rektoratsrede* at the University of Bonn, suggested that the forces of crystal structure are identical with chemical valence forces and that a close relationship exists between molecules of higher order, i.e. coordination compounds, and crystals.

Alfred Werner possessed a tremendous capacity for visualization and thinking in terms of three dimensions, and he spoke as if he had actually seen atoms. Two years before his coordination theory, in his *Habilitationsschrift*, 'Beiträge zur Theorie der Affinität und Valenz', he proposed that 'affinity is an attractive force acting equally from the center of an atom toward all parts of its spherical surface' (Werner, 1891). From this definition he concluded that 'separate valence units do not exist' and that valence is 'dependent not upon one atom alone but simultaneously upon the nature of all elementary atoms which combine to form the molecule'. In view of Werner's penchant for thinking in three-dimensional geometric terms, his ideas of valence just quoted and his concepts of coordination numbers and their associated geometric configurations, it is surprising that he did not apply his coordination theory directly to the domain of crystallography. Yet he did not.

For example, Werner apparently did not realize that the polynuclear complexes which he investigated so extensively (Kauffman, 1973b) constitute a transition between the usual mononuclear coordination compounds and the infinite structure of the crystal lattice. Because he knew that certain groups, especially hydroxide, can coordinate with two metal atoms simultaneously to form bridges, it is possible that he might have considered the possibility of infinite structures with metal atoms bonded in this manner. In the case of μ-hydroxo (*ol*) bridges, this infinite olation process will result in the lattice structure of crystalline metal hydroxides, e.g. hydrargyllite, as later established by X-ray crystallography. However, Werner did not reach such conclusions, and it remained for Paul Pfeiffer (1875–1951) (Kauffman, 1974g), Paul Niggli (1888–1953) and others to point out that crystal structures are in beautiful agreement with his coordination theory, as revealed by the then new experimental technique of X-ray diffraction* (Kauffman, 1973e, 1973f).

In 1915 Pfeiffer suggested that crystals be regarded as extremely high-molecular-weight coordination compounds, in which atoms act as coordination centers, about which further atoms group themselves in definite symmetrical relationships (Pfeiffer, 1915). According to him, crystals are constructed according to the same structural chemical and steric laws as coordination compounds. He regarded the forces holding together the atoms or groups of atoms in crystals as identical with the chemical forces operative in coordination compounds. He thus extended the coordination theory into areas in which it had previously been inapplicable.

Pfeiffer dealt with sodium chloride, which he regarded as a high-molecular-weight coordination compound $(NaCl)_n$ made up of equal amounts of $[NaCl_6]$ and $[ClNa_6]$ units. He showed that in crystals of symmetrical compounds, the difference between primary valencies (*Hauptvalenzen*) and secondary valencies (*Nebenvalenzen*) disappears. Pfeiffer extended his treatment to other crystals such as the diamond, zinc blende (ZnS), fluorite (CaF_2), copper, silver, intermetallic compounds, anhydrite ($CaSO_4$) and calcite ($CaCO_3$), and he showed that coordination centers can be groups of atoms as well as single atoms (Pfeiffer, 1916). He also pointed out that coordination numbers as high as twelve must sometimes be considered (Pfeiffer, 1918), and he suggested that in crystals of simple organic molecular compounds of type AB each constituent acts as a coordination center so that AB_6 and BA_6 units interpenetrate just as they do in the rock salt crystal (Pfeiffer, 1920). Niggli extended the coordination theory to more complex crystalline compounds.

DETERMINATION OF CONFIGURATION BY X-RAY DIFFRACTION: WYCKOFF AND POSNJAK (1921) AND DICKINSON (1922)

As a result of Paul Pfeiffer's suggestion of applying Werner's coordination theory to crystals and the advent of new experimental techniques (Max von Laue's X-ray diffraction method in 1912, William Lawrence Bragg and William Henry Bragg's method for obtaining the distances between crystal planes in 1912 and Peter Debye and Paul Scherrer's powder method in 1916), a number of scientists in various countries simultaneously began to investigate the crystal structures of coordination compounds by means of X-rays. In the words of Ralph W. G. Wyckoff

Plate 23.   Ralph W. G. Wyckoff (b. 1897, photograph taken in 1930)
[Courtesy, Professor R. W. G. Wyckoff, University of Arizona, Tucson,
Arizona]

(b. 1897), now Professor of Physics and Bacteriology at the University
of Arizona and one of the pioneers in crystallography (Plate 23):

> Werner's theory of coordination must be counted one of the
> great steps forward in our understanding of chemical combina-
> tion. Concerned with the distribution of atoms in molecular
> complexes and coming not long before the discovery of X-ray
> diffraction, it was particularly important for those of us who
> were then beginning crystal analysis. This analysis, in establish-
> ing for the first time exactly where the atoms are in a solid,
> offered the most direct check imaginable of how   correct
> Werner's notions about valence were, and, conversely, the ideas
> about coordination arising from this theory could suggest many
> compounds that it would be profitable to examine with X-rays
> (Wyckoff, 1967).

According to Wyckoff, the X-ray diffraction method also provided 'an
unexpectedly direct way to ascertain the measure of reality behind the
Werner theory and its implied equivalence of some "primary" and
"secondary" bonds'.

Shortly after receiving his doctorate, Wyckoff 'chose ammonium chloroplatinate as a crystal that should provide a clear-cut test of Werner coordination'. The results of his investigation, published together with Eugen Posnjak (1888–1949) as co-author, constitute the first published experimental crystallographic study of a coordination compound (Wyckoff and Posnjak, 1921). In Wyckoff's own words:

All six chlorine atoms in $(NH_4)_2 PtCl_6$ . . . were crystallographically identical. They were equally distant from the metal atom, and hence there was no difference in the bonds they formed with it. Furthermore, chlorines were found to be at the corners of a regular octahedron having the platinum atom at its center. A more complete agreement with the predictions of the Werner theory could scarcely have been imagined (Wyckoff, 1967).

Others quickly applied the X-ray diffraction technique and confirmed the octahedral configuration of the six halogen atoms in similar hexacoordinate complexes, e.g. in 1921 and 1922 alone, $K_2 [PtCl_6]$, $[Co(NH_3)_6]Cl_2$, $[Ni(NH_3)_6]Cl_2$, $Rb_2 [PdBr_6]$, $(NH_4)_2 [SnCl_6]$ and $(NH_4)_2 [SiF_6]$. Wyckoff also confirmed Werner's view that the molecules of water and of ammonia in most crystalline salt hydrates and ammines are associated with the metallic atom in the same way as the coordinated atoms and radicals in a complex anion. Thus he showed that $[Ni(NH_3)_6]Cl_2$ has the same crystal structure as $(NH_4)_2 [PtCl_6]$ with $[Ni(NH_3)_6]^{2+}$ ions in place of $[PtCl_6]^{2-}$ ions. He found the structures of $[Ni(NH_3)_6](NO_3)_2$ and $NiSO_4 \cdot 6H_2O$ to be similar. In the case of $NiSO_4 \cdot 7H_2O$, in 1935 Beevers and Schwartz confirmed Werner's prediction of octahedral coordination of six water molecules around the metal ion, with the seventh water molecule situated elsewhere in the structure, unattached to the metal. Since the early 1920s the structures of numerous coordination compounds of various coordination numbers have been determined by the X-ray diffraction method. To quote Wyckoff once more:

Results such as these have put the basic correctness of the coordination theory beyond dispute; its formulation will remain permanently useful even though crystal structure methods have advanced until we no longer need rely on its predictions (Wyckoff, 1967).

Within a year of Wyckoff and Posnjak's confirmation of the octahedral configuration for platinum(IV) (coordination number six), Roscoe Gilkey Dickinson, Professor of Physical Chemistry at the California Institute of

Plate 24. Roscoe Gilkey Dickinson (1894–1945) [Courtesy, Professor
John D. Roberts, California Institute of Technology, Pasadena, California]

Technology (1894–1945) (Plate 24), used the X-ray diffraction method
to confirm Werner's prediction of a planar configuration for platinum(II)
(coordination number four) (Dickinson, 1922). Dickinson followed up
his determination of the crystal structure of ammonium hexachloro-
stannate(IV) ((NH₄)₂ [SnCl₆]), isomorphous with Wyckoff and Posnjak's
(NH₄)₂ [PtCl₆], with determinations of the crystal structures of
potassium tetrachloroplatinate(II) (K₂ [PtCl₄]) and potassium and ammonium tetrachloropalladates(II) (K₂ [PdCl₄] and (NH₄)₂ [PdCl₄]). In the
three cases each platinum or palladium atom was found to be surrounded
by four equidistant and equivalent chlorine atoms situated in a plane.
A similar arrangement has been found for the ammonia molecules in
[Pt(NH₃)₄]Cl₂.

Since Dickinson's first determinations, crystal structures of many other
complexes of various coordination numbers have been determined. These
determinations have included nonelectrolytes as well as salts. For example, the *cis* and *trans* isomers of [Pt(NH₃)₂Cl₄], upon which Werner
based so great a portion of his original paper (Werner, 1893), have been
shown to be composed of discrete molecules. Other complexes whose
crystal structures have been determined include [Co(NH₃)₃(NO₂)₂Cl],
[Ni((NH₂)₂CS)₄]Cl₂, [Fe(acetylacetonate)₃] and [Ni(C₂H₄(NH₂)₂)₃]-
(NO₃)₂.

All these investigations and others have provided a complete and *direct* confirmation of Werner's views to support his *indirect* configurational proofs obtained during the previous decades by preparation of isomers and resolution of optically active compounds (see Chapter 6), and today the terminology and concepts of coordination theory are routinely used in crystallography. For example, in a compound such as $K_2[PtCl_6]$, three of Werner's structural units (octahedron, cube and tetrahedron) are found; the Pt^{4+} ion is surrounded octahedrally by six chloride ions, forming the $[PtCl_6]^{2-}$ ion, which acts as a coordination center and is surrounded cubically by eight K^+ ions, while the K^+ ions group the $[PtCl_6]^{2-}$ ions tetrahedrally around themselves (Kauffman, 1978a).

THE EFFECTIVE ATOMIC NUMBER CONCEPT: SIDGWICK (1923)

As we have already seen, theories of valence had not advanced sufficiently in Werner's time to enable him to interpret the bonding in coordination compounds and to explain adequately the relationships between his much maligned concepts of principal and auxiliary valence. Nevertheless, Werner's recognition, an entire generation before the views of Kossel and Lewis, of the difference in bonding between the central metal atom and ligands in the inner sphere and that between the central metal atom and groups in the outer sphere probably led to our currently accepted concepts of covalent and ionic bonding. In the words of Gilbert Newton Lewis, who always acknowledged his debt to Werner, 'We must consider Werner's theory of coordination numbers as the most important principle at present available for the classification of polar compounds'.

Recognition of the two basic types of electron interaction — transfer and sharing — did much to dispel one of the great objections to Werner's theory, viz. that while some simple compounds, 'compounds of the first order' (*Verbindungen erster Ordnung*) in Werner's terminology, are ionic such as sodium chloride, others such as methane are not. It soon became clear that Werner's 'compounds of the first order' could themselves be further subdivided into two limiting types — ionic (formed by transfer of electrons from the metal to the nonmetal) and covalent (formed by a sharing of electrons between the atoms). Furthermore, the covalent 'compounds of the first order' were found to possess many properties in common with Werner's complex compounds, i.e. 'compounds of higher order' (*Verbindungen höherer Ordnung*). In other words, one form — the covalent form — of Werner's *Hauptvalenzen* or principal valences was quite similar to his *Nebenvalenzen* or auxiliary valences. This finding was in close agreement with Werner's repeated statements that there was no essential difference between his *Hauptvalenzen* and *Nebenvalenzen* except in their origin and mode of formation. Werner's *Hauptvalenz* is now considered to be identical with what is today known as oxidation number or oxidation state.

The interpretation of Werner's *Nebenvalenz* is somewhat more complicated in terms of electronic structure. From Lewis' covalent model was developed the valence bond (VB) theory of coordination associated with the names of Linus Pauling and John C. Slater. This theory, closely related to theories of hybridization and resonance, was the first successful application of the electronic theory of valence to coordination compounds, and from the early 1930s to the early 1950s virtually all coordination phenomena were interpreted in terms of it. The VB theory gave simple and satisfactory answers to questions of geometry and magnetic susceptibility, with which chemists of that period were concerned.

From Kossel's ionic model was developed the crystal field theory (CFT) of coordination associated with the names of Hans Bethe and John H. Van Vleck. Although his theory was used to some extent by physicists as early as the 1930s, it did not find general acceptance among chemists until the 1950s. When modified to include some degree of covalence, crystal field theory is usually known as ligand field theory (LFT) or adjusted crystal field theory (ACFT) and is currently the best approach to the problem of quantitative treatment of spectra and other properties. Both valence bond (VB) theory and crystal field theory (CFT) are only simplifications of the more general but more complicated molecular orbital (MO) theory, which today offers the best interpretation of the properties of coordination compounds.

Plate 25. Nevil Vincent Sidgwick (1874–1952) [(1958), *Proc. Chem. Soc.* 312]

The first attempts to interpret Werner's views on an electronic basis, however, were made in 1923 by Nevil Vincent Sidgwick, a bachelor don at Lincoln College, Oxford University (1873–1952) (Plate 25), and Thomas Martin Lowry (1874–1936) (Lowry, 1923). Until that year Sidgwick's primary research interest had been what we would today call physical organic chemistry. Probably his most important work up to that time was his recognition of the anomalously low solubility and anomalously high vapor pressure of o-nitrophenol, which, as almost every beginning organic chemistry student knows, led ultimately to the recognition of chelate hydrogen bonding. The electronic interpretation of chemical constitution was an entirely new interest to Sidgwick. It was engendered by his acquaintance with Lord Ernest Rutherford, whom he first met in 1914, and was stimulated by one of Niels Bohr's papers 'Atomic Structure' (Bohr, 1923) as well as by Bohr's book *The Theory of Spectra and Atomic Constitution* (Bohr, 1922) and still further by the influence of Gilbert Newton Lewis, who stayed with Sidgwick in June 1923. As late as 1923 Sidgwick told a pupil who had brought him an inorganic paper to grade, 'I don't know any inorganic chemistry'.

Sidgwick's initial concern was to explain Werner's coordination number in terms of the sizes of the sub-groups of electrons in the Bohr atom (Sidgwick, 1923). He soon developed the attempt to systematize coordination numbers into his concept of the 'effective atomic number' (EAN) (Sidgwick, 1927). He considered ligands to be Lewis bases* which donated electrons (usually one pair per ligand) to the metal ion, which thus behaves as a Lewis acid*. Ions tend to add electrons by this process until the EAN (the sum of the electrons on the metal ion plus the electrons donated by the ligand) of the next noble gas is achieved.

Today the EAN rule is of little theoretical importance. It attributes a particular stability to noble gas configurations without explaining this fact. Furthermore, although a number of elements obey Sidgwick's EAN rule (e.g. coordination number 6: Fe(II), Co(III), Pd(IV), Ir(III), Pt(IV); coordination number 4: Cu(I), Ag(I)), there are many important stable exceptions (e.g. coordination number 6: Cr(III), Fe(III), Ni(II), Ir(IV); coordination number 4: Co(II), Ni(II), Pd(II), Pt(II)). Nevertheless, it is extremely useful as a predictive rule in one area of coordination chemistry, that of metal carbonyls and nitrosyls which obey it with a fairly high frequency. Finally, it is of historical significance in being the result of the first attempt to explain Werner's coordination theory in terms of electronic structure (Kauffman, 1978a).

THE *TRANS* EFFECT: CHERNYAEV (1926)

Almost every beginning student of organic chemistry knows that substitution reactions do not occur in a random manner. For example, according to the Crum Brown–Gibson rule governing the path of substitution reactions in aromatic compounds, certain groups on the benzene nucleus are *ortho-* or *para*-orienting, while others are *meta*-orienting. In a similar manner, substitution reactions among coordination compounds are not random. However, the general principle underlying the directive influences of coordinated ligands was not enunciated until well into the third decade of the present century. Such influences are most pronounced and well investigated among square planar complexes, especially those of platinum(II) (Kauffman, 1976g, 1976h, 1977a).

The chemical behavior of dipositive platinum complexes was studied by many of the early investigators in coordination chemistry, and the well-known regularities observed in substitution reactions were cited by Werner in his assignment of *cis* or *trans* configurations for platinum(II) complexes, to which he ascribed a square planar arrangement (Werner, 1893). The compounds chosen by Werner were among the simplest and longest known platinum isomers, viz. platosemidiammine chloride or Peyrone's Salt and platosammine chloride or Reiset's Second Chloride, both with the formula $Pt(NH_3)_2Cl_2$. According to the Blomstrand–Jørgensen chain theory, these compounds possess the configurations:

$$\underset{\text{Platosemidiammine chloride}}{Pt \begin{array}{l} \diagup NH_3-NH_3-Cl \\ \diagdown Cl \end{array}} \qquad \underset{\text{Platosammine chloride}}{Pt \begin{array}{l} \diagup NH_3-Cl \\ \diagdown NH_3-Cl \end{array}}$$

while Werner, on the basis of transformation reactions, assigned them the configurations:

$$\underset{cis}{\begin{array}{c} Cl \diagdown \quad \diagup NH_3 \\ Pt \\ Cl \diagup \quad \diagdown NH_3 \end{array}} \qquad \underset{trans}{\begin{array}{c} Cl \diagdown \quad \diagup NH_3 \\ Pt \\ H_3N \diagup \quad \diagdown Cl \end{array}}$$

The synthesis of each of these compounds involves directive influences, and the preparative reactions were known as Peyrone's reaction and Jørgensen's reaction, respectively, and were said to exemplify Peyrone's rule (*cis* orientation) and Jørgensen's rule (*trans* orientation):

$$K_2\left[\begin{array}{c}X\diagdown\diagup X\\ \quad Pt\\ X\diagup\diagdown X\end{array}\right] + 2A \longrightarrow \left[\begin{array}{c}A\diagdown\diagup X\\ \quad Pt\\ A\diagup\diagdown X\end{array}\right] + 2KX \qquad \text{Peyrone's reaction}$$

cis

$$\left[\begin{array}{c}A\diagdown\diagup A\\ \quad Pt\\ A\diagup\diagdown A\end{array}\right]X_2 + 2HX \longrightarrow \left[\begin{array}{c}A\diagdown\diagup X\\ \quad Pt\\ X\diagup\diagdown A\end{array}\right] + 2AHX \qquad \begin{array}{l}\text{J\o rgensen's reaction}\\ \text{(A = NH}_3\text{ or an amine, X = halogen.)}\end{array}$$

trans

In 1893, the year of the publication of the coordination theory, a third important regularity was observed by the Russian chemist Nikolaĭ Semenovich Kurnakov (1860–1941) (Kauffman and Beck, 1962). While investigating the substitution of ligands by thiourea and thioacetamide, Kurnakov found that replacement occurs with all the ligands of the *cis* compound but only with the acid radicals of the *trans* compound:

$$\left[\begin{array}{c}A\diagdown\diagup X\\ \quad Pt\\ A\diagup\diagdown X\end{array}\right] + 4tu \longrightarrow \left[\begin{array}{c}tu\diagdown\diagup tu\\ \quad Pt\\ tu\diagup\diagdown tu\end{array}\right]X_2 + 2A$$

cis

$$\left[\begin{array}{c}A\diagdown\diagup X\\ \quad Pt\\ X\diagup\diagdown A\end{array}\right] + 2tu \longrightarrow \left[\begin{array}{c}A\diagdown\diagup tu\\ \quad Pt\\ tu\diagup\diagdown A\end{array}\right]X_2$$

trans

(A = NH₃ or an amine, X = halogen or acid radical, tu = thiourea)

Since the two isomers yield different products, this reaction, known as Kurnakov's reaction or Kurnakov's test, may be used to differentiate *cis* from *trans* isomers of dipositive platinum or palladium. Kurnakov's classic reaction played a crucial role in Werner's proof of the square planar configuration of Pt(II) and in Chernyaev's formulation of the *trans* effect (Kauffman, 1978a).

Peyrone's reaction, J\o rgensen's reaction and Kurnakov's reaction are all specialized cases of a more general directive influence and hence are explainable by and derivable from it. Although Werner recognized the principle of '*trans* elimination' as early as 1893, it was not until 1926 that the Russian chemist and later academician Il'ya Il'ich Chernyaev (1893–1966) (Plate 26) pointed out the general regularity of what he called the *trans* effect in order to describe the influence of a coordinated ligand on the practical ease of preparing compounds in which the group *trans* to it had been replaced (Chernyaev, 1926).

Plate 26. Il'ya Il'ich Chernyaev (1893–1966) [Courtesy, the late Academician Il'ya I. Chernyaev, Director, N. S. Kurnakov Institute of General
and Inorganic Chemistry of the Academy of Sciences of the USSR,
Moscow, USSR]

Chernyaev generalized that a negative group coordinated to a metal
atom loosens the bond of any group *trans* to it and thus explained not
only Peyrone's, Jørgensen's and Kurnakov's reactions but also many other
features of the reactions of divalent and tetravalent platinum (Chernyaev,
1927). He also investigated substitution reactions of complexes of
chromium, cobalt, tellurium and osmium. He postulated that the *trans*
effects of atoms are inversely proportional to their metallic character,
i.e. directly proportional to their electronegativities. Electronegative
ligands such as NO_2^-, NCS^-, F^-, Cl^-, Br^- and I^- have a greater '*trans*
influence*' than neutral ligands such as NH_3, ammines or H_2O. Chernyaev's original *trans*-directing series has been extended to include a variety
of ligands: $CN^- \sim CO \sim C_2H_4 \sim NO \sim H^- > CH_3^- \sim SC(NH_2)_2 \sim$
$SR_2 \sim PR_3 > SO_3H^- > NO_2^- \sim I^- \sim SCN^- > Br^- > Cl^- > C_5H_5N >$
$RNH_2 \sim NH_3 > OH^- > H_2O$.

Chernyaev's *trans* effect has been useful not only in synthetic work but
also in structure-proof. His discovery enabled him and his many students
and research workers to prepare numerous complexes not only of
platinum but also of palladium, rhodium, iridium, ruthenium, cobalt and

other metals. The rule made it possible for the first time to plan systematic routes for carrying out inner-sphere substitution reactions in order
to prepare platinum complexes in which all the ligands are different.
For example, Chernyaev's synthesis of the three possible geometric
isomers of $[Pt(NH_3)(C_5H_5N)(NH_2OH)(NO_2)]^+$ was cited as evidence for
a square planar arrangement (see p. 165) (Chernyaev, 1928). In another
application of the *trans* effect, Anna D. Gel'man *et al.* were able to prepare the three possible geometric isomers of bromochloropyridineammine-
platinum(II) (Gel'man *et al.*, 1948, 1949):

$$\begin{bmatrix} C_5H_5N & \!\!\!\!\begin{smallmatrix} Cl \\ \diagdown \\ Pt \\ \diagup \end{smallmatrix}\!\!\!\! & \begin{smallmatrix} Br \\ \diagup \\ \\ \diagdown \end{smallmatrix} \\ & & NH_3 \end{bmatrix} \quad \begin{bmatrix} H_3N & \!\!\!\!\begin{smallmatrix} Cl \\ \diagdown \\ Pt \\ \diagup \end{smallmatrix}\!\!\!\! & \begin{smallmatrix} Br \\ \diagup \\ \\ \diagdown \end{smallmatrix} \\ & & NC_5H_5 \end{bmatrix} \quad \begin{bmatrix} C_5H_5N & \!\!\!\!\begin{smallmatrix} Cl \\ \diagdown \\ Pt \\ \diagup \end{smallmatrix}\!\!\!\! & \begin{smallmatrix} NH_3 \\ \diagup \\ \\ \diagdown \end{smallmatrix} \\ & & Br \end{bmatrix}$$

They were also able to accomplish a similar synthetic *tour de force* for
platinum(IV) by preparing several of the fifteen possible isomers of a compound of the MABCDEF type, viz. $[Pt(NH_3)(C_5H_5N)(Cl)(Br)(NO_2)I]$
(Essen and Gel'man, 1956; Essen *et al.*, 1958).

Chernyaev's concept is one of the fundamental principles of synthetic
inorganic chemistry and has greatly stimulated the theoretical study of
the reactivity and kinetics of coordination compounds, and various
chemists have sought to define it more precisely. 'The *trans* effect or *trans*
influence of a group coordinated to a metal ion is the tendency of that
group to direct an incoming group into the *trans* position to itself' (Chatt
et al., 1955). 'The *trans* effect stipulates that the bond holding a group
trans to an electronegative or otherwise labilizing group is weakened'
(Quagliano and Schubert, 1952). 'In compounds with square or octahedral
structure with a central complex-forming cation, the rate of substitution
of an atom or molecule linked to the central atom is determined by the
nature of the substituent at the opposite end of the diagonal. Thus the
stability of the bond between this (central) atom and any substituent is
little affected by the character of the neighboring atoms or molecules, but
is greatly influenced by those more distant, in the *trans* position, on the
diagonal of the square' (Chernyaev, 1957). The *trans* effect is 'the effect
of a coordinated group upon the *rate* of substitution reactions of ligands
opposite to it. Metals in which the rate influence of opposite, or *trans*
groups, is definitely greater than the influence of adjacent, or *cis* groups,
will be considered to show a *trans* effect' (Basolo and Pearson, 1962).

Although the Russian word *transvliianie* has been translated as either
'*trans* effect' or '*trans* influence', some modern authors make a distinction
between the two phrases. Thus, the *trans effect* is concerned with
kinetics* and the *rate* of substitution reactions, as in Basolo and Pearson's

definition. On the other hand, the *trans influence* is concerned with thermodynamics* and has been defined as 'the extent to which that ligand weakens the bond *trans* to itself in the equilibrium state of a substrate'. The distinction is of value in connection with the two theoretical viewpoints concerning the possible mechanism of the *trans* effect.

The first type of theory is primarily an electrostatic one that emphasizes a weakening or labilization of the *trans* bond. The first attempt at such an explanation was made by Chernyaev himself, who recognized that a simple Coulombic explanation was inadequate (Chernyaev, 1927). Another early electrostatic theory was advanced in 1935 by B. V. Nekrasov. Building upon Chernyaev and Nekrasov's ideas, Aleksandr Abramovich Grinberg noted that the *trans* effect for the series OH^-, Cl^-, Br^- and I^- increases in an order corresponding to their increase in molar refraction, a fact which suggested that Kasimir Fajans' concept of polarizability is pertinent here (Grinberg, 1932). Grinberg's polarization theory is the earliest theory of the *trans* effect that still has current application.

The second type of theory emphasizes the lowering of the activation energy of *trans* replacement and makes use of modern molecular orbital theory (Chatt *et al.*, 1955). According to this line of reasoning, two π-bonding ligands competing for the *d* orbitals of the metal tend to labilize each other, compared to the more stable *cis* isomer where no competition takes place, and the stronger π-bonder will weaken the bonding of the ligand *trans* to it. Several interpretations have also recently been made to explain the *trans* effect on the basis of σ-bonding only. What interpretation of the *trans* effect is the best is currently uncertain.

PROOF OF THE SQUARE PLANAR CONFIGURATION FOR PLATINUM(II): MILLS AND QUIBELL (1935)

The proof for the octahedral configuration of cobalt(III) and platinum(IV) (coordination number six) as well as the exclusion of the three other theoretically possible configurations (hexagonal pyramidal, hexagonal planar and trigonal prismatic) came from three main lines of evidence: (1) chemical evidence such as 'isomer counting' and transformation reactions (see pp. 105–136); (2) resolution of selected compounds (Werner, 1911, 1914); and (3) X-ray diffraction studies (see pp. 151–155). The first two lines of evidence were pursued successfully largely through the efforts of Alfred Werner.

The proof for the square planar configuration of platinum(II) came from the same three lines of evidence, but with a different chronology and a difference in the interpretation of the resolution experiments. The three theoretically possible symmetrical configurations for coordination number four are:

(1) Tetrahedral   (2) Square planar (3) Square pyramidal

The method is summarized in Table VI.

The tetrahedral configuration (sp^3) is known for many nontransition elements (Kauffman, 1973a). For example, the well-known tetrahedral configuration for carbon, the first spatial configuration to be experimentally established for any element, has been proven by the absence of isomers of compound types MA_2B_2 and MA_2BC and by the numerous resolutions of compounds of type MABCD. Among metals, resolutions of tetrahedral compounds of type MABCD are rare because such compounds are usually labile and racemize rapidly. Compounds of type $M(\overline{AB})_2$, however, have been resolved, e.g. bis(benzoylacetonato)beryllium. Tetrahedral structures are formed by large ligands and three types of small metal ions: (1) those with a noble gas configuration, ns^2np^6 (e.g. Be^{2+}, Al^{3+}); (2) those with a pseudo-noble gas configuration, $ns^2np^6(n-1)d^{10}$ (e.g. Zn^{2+}, Cd^{2+}, Ga^{3+}); and (3) transition metal ions that do not favor other structures because of crystal field stabilization energy (e.g. Co^{2+}). The square pyramidal configuration has never been observed experimentally and, although theoretically possible, is considered unlikely and is therefore usually eliminated from active consideration in stereochemical work.

The square planar configuration (dsp^2 hybridization, in terms of modern orbital theory) for platinum(II) was proposed by Werner in his very first paper on the coordination theory on the basis of chemical evidence ('isomer counting' and transformation reactions) (Werner, 1893). He also proposed which isomers are *cis* and which are *trans* by correlating their structures with their chemical behavior by a concept that he called '*trans* elimination'. Since only few platinum(II) and palladium(II) isomers were known and similar isomerism among other elements was not discovered for a number of years, Werner's square planar configuration was questioned more and more strongly as time went by, and other grounds were adduced to account for the structures of these compounds. For example, several workers claimed to have isolated more than two isomers for compounds of type MA_2B_2, a fact which would eliminate a square planar configuration even though it is also not compatible with either the tetrahedral or square pyramidal arrangements. Furthermore, others claimed to have resolved tetracoordinate platinum(II) complexes of type $M(\overline{AB})_2$, a fact inexplicable by the square planar configuration.

As we have seen, Roscoe Gilkey Dickinson used X-ray diffraction to prove a square planar arrangement of chloride ions around the central

TABLE VI
Proof of configuration for coordination number four by 'isomer counting' (Kauffman, 1978a)

Compound type	Theoretically predicted isomers			Conclusions
	I Tetrahedral	II Square planar (a special case of III)	III Square pyramidal	
MA₄	One form only	One form only	One form only	None possible
MA₃B	One form only	One form only	One form only	None possible
MA₂B₂	One form only	Two geometric (1,2 cis; 1,3 trans)	Two geometric (1,2 cis; 1,3 trans)	More than one isomer would eliminate I and prove II or III.
MA₂BC	One form only	Two geometric (1,2 cis; 1,3 trans)	Two geometric (1,2 cis; 1,3 trans). Cis is asymmetric.	More than one isomer would eliminate I. Resolution of cis isomer would prove III.
MABCD	One asymmetric pair	Three geometric (1,2,3,4; 1,2,4,3; 1,3,2,4)	Three geometric (1,2,3,4; 1,2,4,3; 1,3,2,4). Each consists of an asymmetric pair.	Optical isomerism without geometric isomerism would prove I. Geometric isomerism without optical isomerism would prove II. Both optical isomerism and geometric isomerism would prove III.
M(AĀ)₂[a]	One form only	One form only	One form only	None possible
M(AB̄)₂[b]	One asymmetric pair	Two geometric (1,2,4,3 cis) (1,2; 3,4 trans)	Two geometric (1,2,4,3 cis) (1,2,3,4 trans). Trans is asymmetric.	Optical isomerism without geometric isomerism would prove I. Geometric isomerism without optical isomerism would prove II. Both optical isomerism and geometric isomerism would prove III.
M(AĀ)[a]CD	One form only	One form only	One asymmetric pair	Two isomers would indicate that both I and II exist. Resolution would indicate III.
M(AB̄)[b]CD	One asymmetric pair	Two geometric	Two geometric. Each consists of an asymmetric pair.	One resolvable form would indicate I. Two nonresolvable isomers would indicate II. Two resolvable isomers would indicate III.

[a] $\overline{AA}$ represents a symmetrical bidentate (chelate) ligand, which can span cis positions but not trans positions.

[b] $\overline{AB}$ represents an unsymmetrical bidentate (chelate) ligand, which can span cis positions but not trans positions.

metal atom in $K_2[PtCl_4]$, $K_2[PdCl_4]$ and $(NH_4)_2[PdCl_4]$ (Dickinson, 1922). Additional X-ray structural studies of other compounds confirmed Dickinson's results. Linus Pauling explained the planar structure of platinum(II) and palladium(II) compounds on a theoretical basis and predicted a similar configuration for diamagnetic compounds of nickel(II), gold(III), copper(III) and silver(III).

Since chemical methods for determining configurations depend upon retention of configuration during the various transformation reactions, platinum and its congeners, many of whose complexes are nonlabile, are almost ideally suited for such studies. Furthermore, as we have just seen (pp. 158–162), the orienting effect of ligands in substitution reactions in square planar platinum(II) complexes (the *trans* effect), has been of inestimable value in preparing platinum complexes of known structure. Thus Il'ya Il'ich Chernyaev, discoverer of the *trans* effect, used this regularity of behavior to verify the square planar configuration for dipositive platinum. As can be seen from Table VI, a compound of type MABCD should theoretically exist in three geometrically isomeric forms. Chernyaev (1928) prepared the three possible isomers:

The occurrence of three geometric isomers conclusively eliminated the tetrahedral arrangement but did not absolutely rule out the square pyramidal configuration. It could still be logically argued that each of Chernyaev's isomers consists of an asymmetric pair, but that he was unable to resolve them. It is, of course, philosophically impossible to prove a negative proposition. In other words, another problem of 'negative' evidence similar to that which Werner had faced (see p. 121) had again arisen.

Harry D. K. Drew *et al.* (1934, 1937) attempted to solve the problem by preparing isomers of compound types M($\overline{AA}$)CD and M($\overline{AB}$)CD (see Table VI). For the symmetrical chelate group he chose ethylenediamine, and for the unsymmetrical one he chose isobutylenediamine (1,2-diamino-2-methylpropane). As an example of the first type complex he obtained one and only one form of:

For the second type complex he obtained the two geometric isomers:

$$\begin{bmatrix} C_2H_5NH_2 & NH_2CH_2 \\ & Pt & \\ H_3N & NH_2C(CH_3)_2 \end{bmatrix} Cl_2 \quad \begin{bmatrix} C_2H_5NH_2 & NH_2C(CH_3)_2 \\ & Pt & \\ H_3N & NH_2CH_2 \end{bmatrix} Cl_2$$

Drew's experimental data are in accord with the square planar configuration. He did not consider the square pyramidal configuration. At first sight, it would appear as if a successful resolution would be of no value in proving the square planar configuration. In fact, we have already seen that alleged resolutions of compounds of type $M(\overline{AB})_2$ had been used in attempts to disprove this configuration. A generation before, Werner had realized that successful resolution of coordination compounds would offer a 'positive' proof that hexacoordinate cobalt(III) possesses an octahedral configuration. Similarly, the English stereochemist William Hobson Mills (1873–1959) (Plate 27) realized that resolution of a certain type of bisbidentate complex of platinum(II) would permit a definite decision to be made between the planar and tetrahedral configurations. Mills and Quibell (1935) prepared dichloro(1,2-diamino-2-methylpropane)-platinum(II) by the action of isobutylenediamine on potassium tetra-chloroplatinate(II):

$$K_2[PtCl_4] + (CH_3)_2C(NH_2)CH_2NH_2 \rightarrow \begin{bmatrix} & CH_2-NH_2 & Cl \\ H_3C & & Pt \\ H_3C & C-NH_2 & Cl \end{bmatrix} + 2KCl$$

Plate 27.   William Hobson Mills (1873–1959) [(1960), *Biog. Mem. Fellows Roy. Soc.*, Vol. 6, Courtesy, Professor F. G. Mann, Cambridge University, Cambridge, England]

By the action of *meso*-1,2-diphenylethylenediamine (stilbenediamine) on this compound they obtained a mixed tetraammine salt, 1,2-diamino-2-methylpropane(*meso*-1,2-diamino-1,2-diphenylethane)platinum(II) chloride:

In this compound, if the arrangement of the bonds around the platinum atom were tetrahedral, the chelate rings would lie in planes perpendicular to each other, a configuration which would possess a median plane of symmetry and which hence could not give rise to optical isomerism. On the other hand, if the two chelate rings were coplanar, the cation would be disymmetric, and optical isomerism would be possible.

Mills and Quibell were able to resolve the compound into stable enantiomorphs* by use of diacetyl-*d*-tartaric anhydride as the resolving agent. Since simple complexes such as [en Pt bn]Cl₂ (en = ethylenediamine and bn = isobutylenediamine) have never been resolved, there is no evidence in favor of the improbable square pyramidal configuration. Mills and Quibell's resolution therefore not only effectively eliminated the tetrahedral configuration but also presumably offered convincing proof for the planar configuration of tetracoordinate platinum. Four years later, in what was Mills' last published experimental investigation, Lidstone and Mills (1939) resolved the corresponding palladium(II) compound by the same method. At the time of the publication of these results, other stereochemical and crystallographic evidence for the configuration of platinum (II) and palladium(II) existed, but Mills' resolutions afforded ingenious, elegant and entirely independent proofs (Kauffman, 1978a).

8

Epilogue

Since World War II, the once relatively neglected field of inorganic chemistry has undergone a resurgence of interest and activity and has attracted the attention of more and more scientists. This so-called 're-naissance in inorganic chemistry' has naturally included coordination chemistry. The majority of metal ions in solution are hydrated, and consequently, as Fred Basolo and Ralph Pearson have pointed out, 'coordination chemistry includes the greater part of all inorganic chemistry'. In 1972, according to James E. Huheey, 'a survey of articles in recent issues of the journal *Inorganic Chemistry* indicates that perhaps 70% could be considered to deal with coordination compounds'.

In his 1965 book on the history of the theory of valency, W. G. Palmer declares that it is 'yet too early to assess the very rapid developments since 1930 in a just historical perspective'. The same statement can be made about the history of coordination chemistry. Consequently, for this reason and because of space limitations, our detailed account must close at this point. In general, research, discoveries and innovations in the field since the 1930s have taken place at an ever-accelerating pace.

Classical coordination chemistry, despite the guiding star of Werner's coordination theory, was still largely empirical. Modern coordination chemistry has been characterized by the introduction of and increased emphasis upon comprehensive theories of chemical bonding that have served to integrate and elucidate the immense amount of experimental data. As we have seen, the valence bond (VB) theory of coordination associated with the names of Linus Pauling and John C. Slater was developed from Gilbert Newton Lewis' covalent model (1916). This theory, closely related to theories of hybridization and resonance, constituted the first successful application of the electronic theory of valence to coordination compounds, and from the early 1930s to the early 1950s virtually all coordination phenomena were interpreted in terms of it.

It gave simple and satisfactory answers to questions of geometry and magnetic susceptibility, with which chemists of that period were concerned.

Walther Kossel's ionic model (1916) was revitalized and developed into the crystal field theory (CFT) of coordination associated with the names of Hans Bethe and John H. Van Vleck. Although used to some extent by physicists as early as the 1930s, this theory did not find general acceptance among chemists until the 1950s. When modified to include some degree of covalence, crystal field theory is usually known as ligand field theory (LFT) or adjusted crystal field theory (ACFT) and is currently the best and most popular approach to the problem of quantitative treatment of spectra and other properties. Another simpler electrostatic theory is the valence-shell electron-pair repulsion (VSEPR) theory of directed valency, proposed by R. J. Gillespie and the late R. S. Nyholm in 1957. Both valence bond (VB) theory and crystal field theory (CFT) are only simplifications of the more general but more complicated molecular orbital (MO) theory, which today offers the best interpretation of the properties of coordination compounds. Undoubtedly, the future will see modifications of and changing emphases on these various theories or even the rise of a totally new theoretical viewpoint.

Another characteristic of the newer coordination chemistry is the increasing reliance upon physicochemical methods unknown to Alfred Werner and his contemporaries. Simultaneously with the introduction of these newer techniques, emphasis has shifted from a preoccupation with qualitative studies of structure and stereochemistry to quantitative studies of thermodynamics, kinetics and reaction mechanisms. Some of the areas of current research interest include unusual ligands, oxidation states and coordination numbers; solid state chemistry; photochemistry; relationship between structure and reactivity; variable oxidation state chelates; heteropoly complexes; organometallic compounds such as metallocenes, and π-aromatic complexes with 'sandwich structures'; compounds with metal–metal bonds (metal clusters); clathrates; fluxional coordination compounds; borane complexes; macrocyclic and stereochemically non-rigid ligands; and nitrogen- and oxygen-containing complexes. Many biologically active compounds are complexes, and even the simpler types of complexes have served as model compounds in investigating bodily processes. In fact, the newly emerging field of bioinorganic chemistry is concerned largely with coordination compounds.

In a modest attempt to bring our account up to date, the author has included as Appendix A a chronological list of some historically significant events in coordination chemistry. The word 'some' in the title is used advisedly, for the list is provisional and very incomplete. It is illustrative rather than exhaustive, and no inferences should be drawn as to the relative importance of events included or not included in the list. The author apologizes to the numerous workers in the field whose valuable contributions could not be included because of lack of space.

It is difficult if not impossible to predict the future of coordination chemistry, an exciting field of research in which the solution to any given problem usually opens up a number of new research avenues and poses newer and more challenging problems. If its past history gives any indication of its future course, the continued success, expansion and vitality of coordination chemistry remain assured.

Some historically significant events in coordination chemistry

Known since antiquity Mentioned by Herodotus (*c.* 450 BC)		Alizarin
1597	Andreas Libavius	Tetraamminecopper(II) ion $[Cu(NH_3)_4]^{2+}$
1704	Diesbach	Prussian blue $KFe[Fe(CN)_6]$
1798	'Citizen' Tassaert	Hexaamminecobalt(III) ion $[Co(NH_3)_6]^{3+}$
1811	Jöns Jacob Berzelius	Dualistic electrochemical theory
1813	Louis-Nicolas Vauquelin	Vauquelin's Salt $[Pd(NH_3)_4][PdCl_4]$
1822	Leopold Gmelin	Hexaamminecobalt(III) oxalate $[Co(NH_3)_6]_2(C_2O_4)_3$
1822	Leopold Gmelin	Potassium hexacyanoferrate(III) $K_3[Fe(CN)_6]$
1822	Leopold Gmelin	Hexacyanocobaltates(III) $M_3[Co(CN)_6]$
1822	Leopold Gmelin	Tetracyanoplatinates(II) $M_2[Pt(CN)_4]$
1827	William Christoffer Zeise	Zeise's Salt $K[Pt(C_2H_4)Cl_3]\cdot H_2O$
1837	Thomas Graham	Ammonium theory
1838	J. Gros	Gros' Salt *trans*-$[Pt(NH_3)_4Cl_2]Cl_2$
1840	Jules Reiset	Reiset's First Chloride $[Pt(NH_3)_4]Cl_2$
1841	Jöns Jacob Berzelius	Conjugate theory
1844	Jules Reiset	Reiset's Second Chloride *trans*-$[Pt(NH_3)_2Cl_2]$
1844	M. Peyrone	Peyrone's Salt *cis*-$[Pt(NH_3)_2Cl_2]$

1847	Frederick Augustus Genth	Aquapentaamminecobalt(III) and hexaamminecobalt(III) salts $[Co(NH_3)_5 H_2 O] X_3$ and $[Co(NH_3)_6] X_3$
1850	Charles Gerhardt	Gerhardt's Salt $trans\text{-}[Pt(NH_3)_2 Cl_4]$
1851	Frédéric Claudet	Chloropentaamminecobalt(III) salts $[Co(NH_3)_5 Cl] X_2$
1852	Edmond Fremy	Color nomenclature
1852	O. Wolcott Gibbs	Nitropentaamminecobalt(III) salts $[Co(NH_3)_5 NO_2] X_2$
1853	Wilhelm Hittorf	Transference number determination — first method of studying complexes in solution
1856	O. Wolcott Gibbs and Frederick Augustus Genth	*Researches on the Ammonia–Cobalt Bases*
1856	Carl Ernst Claus	Theory of metal–ammines
1857	O. Wolcott Gibbs	$trans\text{-}[Co(NH_3)_4 Cl_2] Cl$ (Ammonia-praseo salt)
1863	Discovered by J. Morland (1860); investigated by A. Reinecke (1863)	Reinecke's Salt NH_4 $trans\text{-}[Cr(NH_3)_2 (SCN)_4]$
1866	O.L. Erdmann	Erdmann's Trinitrite; Gibbs' Salt $[Co(NH_3)_3 (NO_2)_3]$
1866	Birth of Alfred Werner (12 December, Mulhouse, France)	
1869	Christian Wilhelm Blomstrand	Chain theory of metal–ammines
1870	Per Theodor Cleve	Cleve's Salt $cis\text{-}[Pt(NH_3)_2 Cl_4]$
1878	Sophus Mads Jørgensen	Jørgensen begins work on coordination compounds
1889	Sophus Mads Jørgensen	$trans\text{-}[Co(en)_2 Cl_2] Cl$ (Praseo salts)
1890	Sophus Mads Jørgensen	$cis\text{-}[Co(en)_2 Cl_2] Cl$ (Violeo salts)
1890	Ludwig Mond, C. Langer and F. Quincke	Nickel tetracarbonyl (the first metal carbonyl)
1891	Alfred Werner	*Beitrag zur Theorie der Affinität und Valenz*
1893	Alfred Werner	Coordination theory
1893– 1896	Alfred Werner and Arturo Miolati	Conductivities of coordination compounds
1893	Nikolaï Semenovich Kurnakov	Kurnakov's Reaction (to distinguish *cis* from *trans* isomers of Pt(II))
1897	Alfred Werner	Nomenclature of coordination compounds
1904	Alfred Werner	*Lehrbuch der Stereochemie*
1904	Heinrich Ley; Giuseppe Bruni and C. Fornara	Inner complexes
1905	Alfred Werner	*Neuere Anschauungen auf dem Gebiete der anorganischen Chemie*

1905	H. Grossmann	Use of high ionic strength medium for studying complex constants
1905	Lev Aleksandrovich Chugaev	Dimethylglyoximate test for nickel (first organic spot test reagent for a metal ion)
1907	L. A. Chugaev and V. Sokolov	First coordination compound containing an asymmetric ligand; stereospecificity
1907	Alfred Werner	cis-$[Co(NH_3)_4Cl_2]Cl$ (Ammonia-violeo salts)
1907	Lev Aleksandrovich Chugaev	Cyclic bonding and stability
1908	Marcel Delépine	Metal dithiocarbamates
1908	John Albert Newton Friend	Criticism of Werner's theory
1911	Alfred Werner (with Victor L. King)	Resolution of cis-$Co(en)_2(NH_3)Cl]X$ (asymmetric cobalt atom)
1913	Alfred Werner	Nobel Prize in chemistry
1914	Alfred Werner	Resolution of

$$\left[Co \left\{ \begin{matrix} HO \\ \\ HO \end{matrix} \!\!\! Co(NH_3)_4 \right\}_3 \right] Br_6$$

(first resolution of a completely inorganic coordination compound)

1915	L. A. Chugaev and N. A. Vladimirov	Chugaev's Salt $[Pt(NH_3)_5Cl]Cl_3$
1915	Paul Pfeiffer	Crystals as molecular compounds
1916	Gilbert Newton Lewis	Covalent bonding
1916	Walther Kossel	Ionic bonding
1916	John Albert Newton Friend	*A cyclic theory of the constitution of metal–ammines and of ferro- and ferricyanides*
1916	Walther Kossel	First calculation of energy of a complex by electrostatic model
1918	Hans Hürlimann	Stereospecificity
1919	Fritz Hein	First arene complex
1919	Death of Alfred Werner (15 November, Zürich, Switzerland)	
1921	Ralph W. G. Wyckoff and Eugen Posnjak	Crystal structure of $(NH_4)_2[PtCl_6]$ by X-ray diffraction
1921	Marcel Delépine	Active racemates
1922	Roscoe Gilkey Dickinson	Crystal structure of $K_2[PtCl_4]$ by X-ray diffraction
1923	Nevil Vincent Sidgwick	Effective atomic number concept
1925	Paul Job	Determination of stability constants by method of continuous variations
1925	Il'ya Il'ich Chernyaev	Synthesis of three isomers of $[Pt(NH_3)(NH_2OH)(py)(NO_2)]^+$ (additional proof for square planar configuration for Pt(II))
1926	Johannes Nicolaus Brønsted	S_{N1}CB mechanism (substitution, nucleophilic, first order, conjugate base)

1926	Il'ya Il'ich Chernyaev	The *trans* effect
1927–1931	E.U. Condon, W. Heitler, F. London, Linus Pauling and J.C. Slater	Valence bond theory and orbital hybridization
1929–1932	Hans Bethe (1929), R.S. Mulliken (1932) and John H. Van Vleck (1932)	Crystal field theory (ligand field theory)
1931	Paul Pfeiffer	The Pfeiffer effect
1932	Jannik Bjerrum	Square pyramidal $[Cu(NH_3)_5]^{2+}$
1933	P. Pfeiffer, E. Breith, E. Lubbe and T. Tsumaki	Oxygen-carrying chelate bis(salicylal)ethylenediiminecobalt(II)
1933	Frederick G. Mann	Resolution of Na *cis*-$[Rh(H_2O)_2(SO_2(NH_2)_2)_2]$ (second completely inorganic coordination compound to be resolved)
1934	John C. Bailar Jr and Robert W. Auten	Optical inversion in reactions of cobalt complexes
1934	Reginald P. Linstead	Iron(II) and copper(II) phthalocyanines
1935	K. A. Jensen	Dipole moments to determine structures of Pt(II) isomers
1935	William Hobson Mills and Thomas H. H. Quibell	Resolution of *meso*-stilbenediamino*iso*-butylenediaminoplatinum(II) salts (proof of the square planar configuration for Pt(II))
1937	H. A. Jahn and Edward Teller	Jahn–Teller Effect
1938	T. Tsumaki	Explanation of oxygen-carrying chelates
1938	R. Tsuchida	Spectrochemical series of ligands
1939	Linus Pauling	*The nature of the chemical bond*
1940	John H. Van Vleck and R. Finkelstein	First application of electrostatic model to absorption bands of the ruby
1940	Nevil Vincent Sidgwick and H.M. Powell	Non-bonding pairs of electrons and stereochemistry
1941	Jannik Bjerrum	Theory of the reversible step reactions
1946	Gerold Schwarzenbach	Ethylenediaminetetraacetate in complexometric titrations
1948	H. Irving and R. J. P. Williams	Stability order of complexes
1950	L. N. Essen and A. D. Gel'man	Synthesis of $[Pt(I)(Br)(NH_3)(NO_2)(Cl)(py)]$ (only example of MABCDEF)
1950	J. Chatt and R. G. Wilkins	$Pt(C_2H_4)_2Cl_2$ First example of two double bonds on one metal atom
1951	John W. Irvine Jr and Geoffrey Wilkinson	$[Ni(PX_3)_4]$ (where X = F, Cl or Br)
1951	T. J. Kealey and P. L. Pauson (1951); S. A. Miller, J. A. Tebboth and J. F. Tremaine (1952)	Discovery of biscyclopentadienyliron(II), later called ferrocene
1952	R. B. Woodward, Geoffrey Wilkinson, M. Rosenblum and M. C. Whiting	Recognition of structure, aromatic properties and proposal of name 'ferrocene'

1952	Henry Taube	Inner and outer orbital complexes
1952	Max Wolfsberg and L. Helmholz	Molecular orbital model applied to transition metal complexes
1954	Yukito Tanabe and Satoru Sugano	Calculation of energy level diagrams for octahedral d-group complexes
1954	Geoffrey Wilkinson	Lanthanum cyclopentadienides
1955	E. O. Fischer and W. Hafner	Bisbenzenechromium(0)
1955	Geoffrey Wilkinson and J. M. Birmingham	Biscyclopentadienylrhenium hydride
1955	Y. Saito, K. Nakatsu, M. Shiro and H. Kuroya	Absolute configuration of D- and L-$[Co(en)_3]^{3+}$
1955	J. Chatt, L.A. Duncanson and L.M. Venanzi; L.E. Orgel	π-Bonding theory of the *trans* effect
1956	Dorothy Crowfoot Hodgkin	Crystal structure of the cobalt(III) complex of vitamin B_{12}
1957	Ronald J. Gillespie and Ronald S. Nyholm	Valence-shell electron-pair repulsion theory (VSEPR) of directed valency
1957	Jack Bobinski, Marvin Fein and Nathan Mayes	First carborane (1-isopropenylcarborane)
1958	Claus E. Schäffer and Christian Klixbüll Jørgensen	Nephelauxetic series (interelectronic repulsion) of central atoms and ligands
1958	S. Ahrland, J. Chatt and N. R. Davies	Systematization of donor atom coordination tendencies of metal ions
1959	Elias J. Corey and John C. Bailar Jr	Conformational analysis of coordination compounds
1959	W. H. Zachariason and H. A. Plettinger	Hexagonal bipyramidal 8-coordination, $[UO_2(OCOCH_3)_3]^-$
1959	F. G. A. Stone	First cyclooctatetraene complexes
1959	Leo H. Sommer	First resolution of optically active silicon compounds
1960	Anthony R. Pitochelli and M. Frederick Hawthorne	Icosahedral $B_{12}H_{12}^{2-}$
1960	Stanley Kirschner	Resolution and structure-proof of a hexa-coordinate silicon complex
1960	Robert B. Woodward *et al.*	Proof of structure of chlorophyll by total synthesis
1960	R. A. Marcus	Outer sphere electron transfer theory
1960– 1967	John C. Bailar Jr, M. J. S. Crespi and John Geldard	Action of biological systems on optically active complexes
1961	Christian Klixbüll Jørgensen	Optical electronegativities from electron transfer spectra
1962	R. E. Sievers, R. W. Moshier and M. L. Morris	Resolution of chromium hexafluoride acetylacetonate by gas chromatography
1962	Neil Bartlett	Synthesis of $Xe[PtF_6]$
1962	Fred Basolo and G. S. Hammaker	Linkage isomerism, nitritopenta-ammines of Rh(III), Ir(III) and Pt(IV)

1963	Lauri Vaska	Reversible reaction of $[IrCl(CO)((C_6H_5)_3P)_2]$ with molecular oxygen
1963	Lawrence F. Dahl *et al.*	First metal carbonyl containing a metal cluster
1963	Ralph G. Pearson	Hard and soft acids and bases
1965	B. Rosenberg, L. Van Camp and T. Krigas	Biological effects of platinum complexes
1965	Richard Eisenberg and J. A. Ibers	Trigonal prismatic coordination compounds
1965	A. D. Allen and C. V. Senoff	First metal complex with molecular nitrogen
1965	Cooper H. Langford and Harry B. Gray	σ-Bonding theory of the *trans* effect
1965	Henry Taube	Proof of inner sphere mechanism $[Co(NH_3)_5Cl]^{2+} + Cr^{2+} \rightarrow [Cr(NH_3)_5Cl]^{2+} + Co^{2+}$
1965	D. L. Kepert	Calculations of shape of 6-coordinate complexes
1965	I. I. Chernyaev, L. S. Korablena and G. S. Muraveiskaya	Optically active complexes with only unidentate ligands
1965	M. F. Hawthorne, D. C. Young and P. A. Wegner	First metallocarborane
1965	F. Albert Cotton and Charles B. Harris	Binuclear metal clusters, $[Re_2Cl_8]^{2-}$
1966	Geoffrey Wilkinson	Homogeneous hydrogenation of olefins by $((C_6H_5)_3P)_3RhCl$ (Wilkinson's catalyst)
1966	Stanley Kirschner	Anti-cancer activity of coordination compounds
1967	Jack M. Williams	Hydrated proton $[H_5O_2]^+$ in $[Co(en)_2Cl_2]Cl \cdot HCl \cdot H_2O$
1967	C. J. Pedersen	Crown ethers (complexes with planar, macrocyclic ligands)
1967	James A. Ibers and Derek J. Hodgson	Structure of $[Ir(NO)(CO)((C_6H_5)_3P)_2Cl]^+$ showing bent $M-NO^-$ bond
1968	Andrew Streitwieser Jr and Ulrich Müller-Westerhoff	Bis(cyclooctatetraene)uranium(IV), in which π-molecular orbitals share electrons with uranium f-orbitals
1969	Dale K. Cabbiness and Dale W. Margerum	Recognition and naming of the macrocyclic effect
1969	D. Brown, J. F. Easey and C. E. F. Rickard	Cubic 8-coordination, $Na_3[PaF_8]$
1969	Eugene E. van Tamelen *et al.*	Nitrogen fixation with titanium(II) alcoholate
1969	James P. Collman	Interconvertibility of linear and bent modes of NO coordination

1970	I. H. Sabherval and Anton B. Burg	Synthesis of $Co(NO)_3$
1970	D. A. Owen and M. Frederick Hawthorne	Carborane—transition metal chelate complexes
1971	Rachel Countryman and W. S. McDonald	Cubic 8-coordination, $[N(C_2H_5)_4][U(SCN)_8]$
1971– 1974	P. W. R. Confield and P. G. Eller (1971) P. H. Davis, R. L. Belford and I. C. Paul (1973) N. C. Baenziger, K. M. Dittemore and J. R. Doyle (1974) J. A. Tiethof, A. T. Hetey and D. W. Meek (1974)	Planar 3-coordination, $[((CH_3)_3PS)_3Cu]ClO_4$, $((C_6H_5)_3P)_2CuBr$, $((C_6H_5)_3P)_2AuCl$ and $((CH_3)_3PS)_3Cu$
1973	James P. Collman and S. R. Winter	First stable formyl complex, $(CO)_4Fe(CHO)^-$
1973	G. L. Simon and Lawrence F. Dahl	First 'naked' phosphorus complex
1974	James P. Collman et al.	Picket fence porphyrin
1974	James L. Dye et al.	Cryptated sodium cation and sodium anion
1974	Lawrence F. Dahl	Platinum carbonyl cluster dianions ('tinker toy' construction) with trigonal prismatic configuration
1974– 1975	N. J. de Stephano, D. K. Johnson and L. M. Venanzi (1974); Isao Mochida, J. Arthur Mattern and John C. Bailar Jr (1975)	Trans-spanning ligands in planar complexes
1975	C. M. Lukehart, G. P. Torrence and J. V. Zerle	Diacetylmetallate anions
1976	E. C. Baker et al.	Lanthanide and actinide organometallics
1976	Loren D. Lower and Lawrence F. Dahl	Cube-shaped metal cluster compound, $Ni_8(CO)_8(PC_6H_5)_6$, the first transition metal analog of cubane
1978	R. D. Gillard and F. L. Wimmer	Third resolution of a completely inorganic complex, $[Pt(S_5)_3]^{2-}$
1979	Satoru Shimba, Shuhei Fujinami and Maraji Shibata	Fourth and fifth resolutions of completely inorganic complexes, cis-cis-cis-$[Co(NH_3)_2(H_2O)_2(CN)_2]^+$ and cis-cis-cis-$[Co(NH_3)_2(H_2O)_2(NO_2)_2]^+$
1979	Alan M. Sargeson	Caged metal ions (macrocycles in three dimensions)
1980	Milenko Ćelap	Resolution of $W(OCOC_5H_4N)_2(ONC_9H_4Cl_2)_2$, the first resolution of an 8-coordinate compound

Appendix B

Glossary

Acid radical. An electronegative atom or group of atoms that does not decompose in ordinary chemical reactions.

Antipode. One of a pair of optical isomers ((+) or (−); *d* or *l*), which are identical in all properties, except those which can be described as right and left, e.g. hemihedral crystal form and direction of rotation of the plane of polarized light. Also called antimer, enantiomer or enantiomorph.

Aquation. Replacement of a ligand (coordinated group) in a complex by one or more water molecules.

Chromatography. Separation of mixtures into their constituents by preferential adsorption on a solid such as a column of silica or a strip of filter paper.

Complex salt. An older synonym for coordination compound, used to distinguish it from a double salt (q.v.), which dissociates into its constituent ions in solution.

Conductivity (conductance). A measure of the ability of a given substance or solution to conduct an electric current, equal to the reciprocal of the resistance.

Conjugate. See COPULA.

Copula. Swedish, *koppling*. Also called conjugate. A term borrowed by Jöns Jacob Berzelius from Charles Gerhardt in an attempt to extricate his electrochemical theory from difficulties. According to Berzelius, 'the active compound conserves its property of uniting with other bodies, while the other term, which we call the copula, has lost all tendency to combination, with certain exceptions'.

Cryoscopic. Pertaining to the determination of the freezing points of liquids or solutions. From the lowering of the freezing point of a solution from that of the pure solvent, the molecular weight of the dissolved solute can be determined.

Doktorand. (German) Candidate for a doctoral degree. Plural, *Doktoranden*; feminine, *Doktorandin*, plural, *Doktorandinnen*.

178

Double salt.	A compound which crystallizes as a single substance but which, on dissolving, dissociates into the constituent ions, in contrast to a complex salt (q.v.). The term is also used in a different sense, e.g. on p. 4 and elsewhere in this book, to designate true complexes of types such as $M_3[CoX_6]$ $(3MX \cdot CoX_3)$.
Enantiomorph.	See ANTIPODE.
Geometric isomerism.	That type of stereoisomerism in which the isomers are not optically active. Also called *cis-trans* isomerism.
Habilitationsschrift.	(German) A research paper required in order to lecture at a European university. There is no exact American equivalent. See *Privat-Dozent.*
Hydrolysis (protolysis).	Decomposition in which a compound or ion is split up into other compounds or ions by reaction with water.
Inversion.	Conversion of an optical isomer into its opposite antipode (q.v.), e.g. (+) to (−) or vice versa. Conversion of a geometric isomer into its opposite isomer, e.g. *cis* to *trans* or vice versa.
Ion.	An electrically charged atom or group of atoms formed by the loss or gain of one or more electrons, as a cation (positive ion), which is created by electron loss and is attracted to the cathode during electrolysis, or as an anion (negative ion), which is created by electron gain and is attracted to the anode during electrolysis.
Isomer.	One of a pair or more of compounds having the same composition but different properties. The term was proposed in 1832 by Berzelius from the Greek ἰσομερής (composed of equal parts).
Isomerization.	A chemical reaction, usually but not always in solution, in which one isomer is converted into another.
Kinetics.	That branch of chemistry dealing with the rate and mechanism of reactions.
Lewis acid.	A substance which can accept a pair of electrons to form a co-ordinate covalent bond, e.g. a transition metal ion.
Lewis base.	A substance which can donate a pair of electrons to form a coordinate covalent bond. Ligands are usually Lewis bases.
Masked.	Hidden or concealed. A masked atom or group is one that is combined in such a manner that its usual properties are subdued.
Molecular asymmetry.	A principle introduced in 1860 by Louis Pasteur, who proposed that optical activity in a substance is caused by an asymmetric arrangement of atoms in the individual molecule.
Molecular rotatory power.	See Rotatory power.
Molecule.	The smallest physical unit, capable of independent existence, of an element or compound, consisting of one or more like atoms in an element and two or more different atoms in

a compound. In contrast to an ion (q.v.), a molecule is electrically neutral.

Mutarotation. A spontaneous change in the optical rotation (rotatory power, q.v.) of a solution.

Optical isomerism. That type of stereoisomerism in which the isomers are non-superimposable mirror images of each other.

Polarimeter. An instrument for quantitatively measuring the rotation of polarized light and hence for detecting the presence of optically active substances (see Plate 18).

Principal or primary valence. Werner's *Hauptvalenz*, identical with Kossel's electrovalence. The valence by which anions are bonded to a metal ion or complex ion.

Privat-Dozent (German) An unsalaried lecturer at a European university whose income is derived from fees paid by the students who enroll in his course.

Professor Nunwiegehts. (German) One of the nicknames given by his students to Werner, who was particularly addicted to the common expression *Nun wie geht's?* (How's it going?).

Racemate. A equimolar mixture of (+)- and (−)-isomers of an optically active compound. More properly called a racemic mixture, the term racemate being reserved for a solid solution of equimolar amounts of the two antipodes (q.v.).

Racemization. The process, usually in solution, whereby one antipode (q.v.) spontaneously loses its optical activity. In the process, an equal quantity of the opposite antipode is formed.

Rotatory dispersion. The ratio of the specific rotations of a substance observed with light of two different wavelengths.

Rotatory power. The rotatory power or optical activity of a dissolved substance depends upon the thickness of the layer traversed by the light, the wavelength of the monochromatic light, the concentration of the solution, the nature of the solvent and the temperature. It is usually reported as specific rotation, $[\alpha]$:

$$[\alpha]_\lambda^t = \frac{\alpha}{dc}$$

where t = temperature (°C), λ = wavelength of light (usually Na D-line, 5896 Å, Hg green, 5461 Å), α = observed optical rotation in degrees (+ if clockwise, − if counterclockwise), d = length of polarimeter tube (dm), and c = concentration of solute (g/ml of solution). A unit which is more useful for comparison between different compounds is the molecular rotation, $[M]$, which is a relative measure of the rotatory power of the compound on a molecular basis:

$$[M]_\lambda^t = \frac{M[\alpha]_\lambda^t}{100}$$

where M = the molecular weight of the substance.

Secondary valence.	Werner's *Nebenvalenz* or auxiliary valence. The valence by which neutral molecules are bonded to a metal ion.
Specific rotatory power.	See Rotatory power.
Stereochemistry.	That branch of chemistry concerned with the spatial arrangement of atoms in molecules or ions. The founding of stereochemistry is generally considered to date from Le Bel and Van't Hoff's proposal of the tetrahedral carbon atom in 1874.
Thermodynamics.	The study, founded on experiment, of the empirical relations between heat energy and other forms of energy.
Valence (valency).	The relative combining capacity of an atom or ion compared with that of the hydrogen atom taken as unity. There are several different kinds of valence, and the term is used rather loosely by chemists.
Walden inversion.	A chemical reaction in which a reversal of the rotatory power of an optically active compound occurs.
Weihnachtskommers.	(German) An annual traditional Christmas party, smoking party and variety show held by Werner's students just before the Christmas recess. Some were quite elaborate and involved the performances of satirical skits and the printing of comical magazines (*Weihnachtskommerszeitungen*) (Kauffman, 1974c).
X-ray diffraction.	The determination of the structure of a crystalline material by means of the diffraction pattern formed when an X-ray passes through it and falls upon a photographic plate.

General bibliography (in chronological order)

1597	Libavius (1597), *Commentationum Metallicarum Libri Quatuor de Natura Metallorum . . .* , vi. p. II, C. 19. Frankfurt.
1710	Anonymous (1710), *Miscellanea Berolinensia* I, 377.
1724a	Woodward, J. (1724), *Philos. Trans. R. Soc. London* **33**, 15.
1724b	Browne, J. (1724), *Philos. Trans. R. Soc. London* **33**, 17.
1798	Tassaert (1798), *Ann. Chim. (Paris)* **28**, 92.
1813	Vauquelin, L.-N. (1813), *Ann. Chim. (Paris)* **88**, 188.
1822a	Gmelin, L. (1822), *Handbuch der theoretischen Chemie*, 2nd edn, p. 928. Varrentrapp, Frankfurt am Main.
1822b	Gmelin, L. (1822), *Schweigger's J. Chem. Phys.* **34**, 325.
1822c	Gmelin, L. (1822), *Schweigger's J. Chem. Phys.* **36**, 230.
1822d	Gmelin, L. (1822), *Handbuch der theoretischen Chemie*, Vol. II, 2nd edn, p. 1692. Varrentrapp, Frankfurt am Main.
1827	Zeise, W. C. (1827), *Pogg. Ann.* **9**, 632; (1831), *Pogg. Ann.* **21**, 497.
1828	Magnus, G. (1828), *Pogg. Ann. Phys. Chem.* **14**, 239.
1837	Graham, T. (1837), *Elements of Chemistry*, J. B. Baillière, London. This book is rare and is better known through its 1840 German translation, *Lehrbuch der Chemie*, transl. by F. J. Otto, Vol. 2, p. 741. F. Vieweg und Sohn, Braunschweig.
1841	Berzelius, J. J. (1841), *Jahresbericht* **21**, 108.
1844	Peyrone, M. (1844), *Ann. Chim. (Paris)* **51**, 1.
1844	Reiset, J. (1844), *Compt. Rend.* **18**, 1100; (1844), *Ann. Chim. (Paris)* [3] **11**, 417.
1851	Claudet, F. (1851), *Philos. Mag.* **2**, 253.
1851	Genth, F. A. (1851), *Keller–Tiedemann's Nordamerikanischer Monatsbericht für Natur- und Heilkunde* **2**, 8.
1852	Fremy, E. (1852), *Ann. Chim. (Paris)* [3] **35**, 257; (1852), *J. Prakt. Chem.* **57**, 95.
1856	Claus, A. (1856), *Ann. Chem.* **98**, 317.
1856a	Gibbs, W. and Genth, F. A. (1856), *Researches on the Ammonia-Cobalt Bases*, Smithsonian Institution, Washington DC.
1856b	Gibbs, W. and Genth, F. A. (1856), *Am. J. Sci.* [2] **23**, 235, 319.

1857a   Gibbs, W. and Genth, F. A. (1857), *Am. J. Sci.* [2] **23**, 248.
1857b   Gibbs, W. and Genth, F. A. (1857), *Am. J. Sci.* [2] **24**, 86.
1864   Kekulé, A. (1864), *Compt. Rend.* **58**, 510.
1869   Blomstrand, C. W. (1869), *Die Chemie der Jetztzeit vom Standpunkt der electrochemischen Auffassung aus Berzelius Lehre entwickelt.* Carl Winter's Universitätsbuchhandlung, Heidelberg.
1875   Gibbs, W. (1875), *Proc. Am. Acad. Arts Sci.* **10**, 1.
1878   Jørgensen, S. M. (1878), *J. Prakt. Chem.* [2] **18**, 209.
1887   Jørgensen, S. M. (1887), *J. Prakt. Chem.* [2] **35**, 417.
1889   Jørgensen, S. M. (1889), *J. Prakt. Chem.* [2] **39**, 1.
1890   Hantzsch, A. and Werner, A. (1890), *Ber.* **23**, 11. For a discussion and annotated English translation see Kauffman, 1966a.
1890a   Jørgensen, S. M. (1890), *J. Prakt. Chem.* [2] **41**, 429.
1890b   Jørgensen, S. M. (1890), *J. Prakt. Chem.* [2] **41**, 440.
1890   Werner, A. (1890), *Ueber räumliche Anordnung der Atome in stickstoffhaltigen Molekülen.* Dissertation, Universität Zürich, 1890.
1891   Werner, A. (1891), *Vierteljahrsschrift der Zürcher Naturforschenden Gesellschaft* **36**, 192. For a discussion and annotated English translation see Kauffman, 1967d, 1967e.
1892   Petersen, J. (1892), *Z. Phys. Chem.* **10**, 580.
1893   Werner, A. (1893), *Z. Anorg. Chem.* **3**, 267. For a discussion and annotated English translation see Kauffman, 1968, pp. 55—88.
1893   Werner, A. and Miolati, A. (1893), *Z. Phys. Chem.* **12**, 35. For a discussion and annotated English translation see Kauffman, 1968, pp. 89—115.
1894a   Jørgensen, S. M. (1894), *Z. Anorg. Chem.* **5**, 147.
1894b   Jørgensen, S. M. (1894), *Z. Anorg. Chem.* **7**, 289.
1894   Werner, A. and Miolati, A. (1894), *Z. Phys. Chem.* **14**, 506. For a discussion and English translation see Kauffman, 1968, pp. 117—139.
1895   Jørgensen, S. M. (1895), *Bull. Acad. R. Sci. Lett. Danemark, Copenhague* 16, footnote.
1897   Jørgensen, S. M. (1897), *Z. Anorg. Chem.* **14**, 404.
1897   Petersen, J. (1897), *Z. Phys. Chem.* **22**, 410.
1897a   Werner, A. (1897), *Z. Anorg. Chem.* **14**, 21.
1897b   Werner, A. and Klein, A. (1897), *Z. Anorg. Chem.* **14**, 28.
1898   Jørgensen, S. M. (1898), *Z. Anorg. Chem.* **16**, 184.
1898   Reitzenstein, F. (1898), *Z. Anorg. Chem.* **18**,152.
1899a   Jørgensen, S. M. (1899), *Z. Anorg. Chem.* **19**, 78.
1899b   Jørgensen, S. M. (1899), *Z. Anorg. Chem.* **19**, 109.
1899   Werner, A. and Vilmos, A. (1899), *Z. Anorg. Chem.* **21**, 145.
1901   Werner, A. (1901), *Ber.* **34**, 2854.
1901   Werner, A. and Herty, C. (1901), *Z. Phys. Chem.* **38**, 331.
1904   Bruni, G. and Fornara, C. (1904), *Att. Accad. Naz. Lincei, Cl. Sci. Fis., Mat., Nat., Rend.* [5] **13**, 11, 26.
1904   Ley, H. (1904), *Z. Elektrochem.* **10**, 954.
1904   Werner, A. (1904), *Lehrbuch der Stereochemie.* Gustav Fischer, Jena.
1905   Werner, A. (1905, 1909, 1913, 1920, 1923), *Neuere Anschauungen auf dem Gebiete der anorganischen Chemie*, F. Vieweg und Sohn, Braunschweig. The

second (1909) edition was translated into English by E. P. Hedley (1911) as *New Ideas on Inorganic Chemistry*, Longmans, Green and Co., London.

1906   Chugaev, L. A. (1906), *Ber.* **39**, 3190.
1906   Werner, A., Bräunlich, F., Rogowina, E. and Kreutzer, C. (1906), in *Festschrift Adolf Lieben zum fünfzigjährigen Doktorjubiläum und zum siebzigsten Geburtstage von Freunden, Verehrern und Schülern gewidmet*, pp. 197–218. C. F. Winter'sche Verlagsbuchhandlung, Leipzig. Reprinted in (1907), *Ann. Chem.* **351**, 65.
1907   Chugaev, L. A. (1907), *J. Prakt. Chem.* [2] **75**, 153; (1907), *J. Prakt. Chem.* [2] **76**, 88.
1907a   Werner, A. (1907), *Ber.* **40**, 4817. For a discussion and English translation see Kauffman, 1968, pp. 141–154.
1907b   Werner, A. (1907), *Chem. News* **96**, 128.
1907c   Werner, A., Berl, E., Zinggeler, E. and Jantsch, G. (1907), *Ber.* **40**, 2103.
1908   Briggs, S. H. C. (1908), *J. Chem. Soc.* **93**, 1564.
1908a   Friend, J. A. N. (1908), *J. Chem. Soc.* **93**, 260.
1908b   Friend, J. A. N. (1908), *J. Chem. Soc.* **93**, 1006.
1908   Werner, A. (1908), *Ber.* **41**, 1062, 2383.
1909   Werner, A. (1909), *Ber.* **42**, 4324.
1911a   Werner, A. (1911), *Ber.* **44**, 1887. For a discussion and English translation see Kauffman, 1968, pp. 155–173.
1911b   Werner, A. (1911), *Ber.* **44**, 3279.
1912   Werner, A. (1912), *Ann. Chem.* **386**, 1.
1913   Werner, A. (1913), *Über die Konstitution und Konfiguration von Verbindungen höherer Ordnung*, Nobel lecture held on 11 December 1913 in Stockholm. Reprinted in German in (1914), *Naturwissenschaften* **2**, 1, in French in (1914), *J. Chim. Phys.* **12**, 133, and in English in (1966), *Nobel Lectures in Chemistry, 1901–1921*, pp. 253–269. Elsevier, Amsterdam.
1914   Werner, A. (1914), *Ber.* **47**, 3087. For a discussion and English translation see Kauffman, 1968, pp. 175–184.
1915   Pfeiffer, P. (1915), *Z. Anorg. Chem.* **92**, 376.
1915   Povarnin, G. (1915), *J. Russ. Phys. Chem. Soc.* **47**, 217, 501, 989, 1787, 2073.
1916   Friend, J. A. N. (1916), *J. Chem. Soc.* **109**, 715.
1916   Pfeiffer, P. (1916), *Z. Anorg. Chem.* **97**, 161.
1917   Briggs, S. H. C. (1917), *J. Chem. Soc.* **111**, 253.
1918   Pfeiffer, P. (1918), *Z. Anorg. Chem.* **105**, 26.
1920a   Pfeiffer, P. (1920), *Z. Anorg. Allg. Chem.* **112**, 81.
1920b   Pfeiffer, P. (1920), *Z. Angew. Chem.* **33**, 37.
1921   Bohr, N. (1921), *Nature (London)* **107**, 104.
1921   Friend, J. A. N. (1921), *J. Chem. Soc.* **119**, 1040.
1921   Wyckoff, R. W. G. and Posnjak, E. (1921), *J. Am. Chem. Soc.* **43**, 2292.
1922   Bohr, N. (1922), *The Theory of Spectra and Atomic Constitution*, The University Press, Cambridge, UK.
1922   Dickinson, R. G. (1922), *J. Am. Chem. Soc.* **44**, 2404.
1923   Lowry, T. M. (1923), *Trans. Faraday Soc.* **18**, 285.
1923   Sidgwick, N. V. (1923), *J. Chem. Soc.* **123**, 725; (1923), *Nature (London)* **111**, 808; (1923), *Trans. Faraday Soc.* **19**, 469; (1923), *Chem. Ind. (London)* **42**. 1203.

1926   Chernyaev, I. I. (1926), *Izv. Inst. Izuch. Plat. Drugikh Blagorodn. Met.* **4**, 243.
1926   Duval, C. (1926), *C. R. Acad. Sci.* **182**, 636.
1927   Chernyaev, I. I. (1927), *Izv. Inst. Izuch. Plat. Drugikh Blagorodn. Met.* **5**, 118.
1927   Sidgwick, N. V. (1927), *The Electronic Theory of Valency*, Chapt. 10. Oxford University Press, London.
1927   Stelling, O. (1927), *Ber.* **60**, 650.
1928   Chernyaev, I. I. (1928), *Izv. Inst. Izuch. Plat. Drugikh Blagorodn. Met.* **6**, 55.
1928   Pfeiffer, P. (1928), *J. Chem. Educ.* **5**, 1098.
1928   Stelling, O. (1928), *Z. Phys.* **50**, 506.
1932   Grinberg, A. A. (1932), *Izv. Inst. Izuch. Plat. Drugikh Blagorodn. Met.* **10**, 47.
1933   Mann, F. G. (1933), *J. Chem. Soc.* 412.
1934   Bailar, J. C. Jr and Auten, R. W. (1934), *J. Am. Chem. Soc.* **56**, 774; Bailar, J. C. Jr (1946), *Inorg. Syn.* 2, 222.
1934   Drew, H. D. K. and Head, F. S. H. (1934), *J. Chem. Soc.* 221.
1935   Mills, W. H. and Quibell, T. H. H. (1935), *J. Chem. Soc.* 839.
1937   Drew, H. D. K., Head, F. S. H. and Tress, H. J. (1937), *J. Chem. Soc.* 1549.
1939   Lidstone, A. G. and Mills, W. H. (1939), *J. Chem. Soc.* 1754.
1942   King, V. L. (1942), *J. Chem. Educ.* **19**, 345.
1948   Gel'man, A. D., Karandashova, E. F. and Essen, L. N. (1948), *Dokl. Akad. Nauk SSSR* **63**, 37.
1949   Gel'man, A. D., Karandashova, E. F. and Essen, L. N. (1949), *Izv. Sektora Plat. Drugikh Blagorodn. Met.* **24**, 60.
1951   Kealy, T. J. and Pauson, P. L. (1951), *Nature (London)* **168**, 1639.
1952   Miller, S. A., Tebboth, J. A. and Tremaine, J. F. (1952), *J. Chem. Soc.* 632.
1952   Quagliano, J. V. and Schubert, L. (1952), *Chem. Rev.* **50**, 201.
1955   Chatt, J., Duncanson, L. A. and Venanzi, L. M. (1955), *J. Chem. Soc.* 4456, 4461.
1956   Essen, L. N. and Gel'man, A. D. (1956), *Zh. Neorg. Khim.* **1**, 2475.
1957   Chernyaev, I. I. (1957), *Zh. Neorg. Khim.* **2**, 475.
1958   Essen, L. N., Zakharova, F. A. and Gel'man, A. D. (1958), *Zh. Neorg. Khim.* **3**, 2654.
1959   Powell, H. M. (1959), *Proc. Chem. Soc.* 73.
1962   Basolo, F. and Hammaker, G. S. (1962), *Inorg. Chem.* **1**, 1.
1962   Basolo, F. and Pearson, R. G. (1962), *Progr. Inorg. Chem.* **4**, 381.
1964   Martin, D. F. and Martin, B. B. (1964), *Coordination Compounds*, pp. 25–26. McGraw-Hill, New York.
1965   Eisenberg, R. and Ibers, J. A. (1965), *J. Am. Chem. Soc.* **87**, 3776; (1966), *Inorg. Chem.* **5**, 411.
1967   Wyckoff, R. W. G., in Kauffman, 1967a, p. 114.
1970   Kuhn, T. S. (1970), *The Structure of Scientific Revolutions*, 2nd edn, University of Chicago Press, Chicago.
1978   Gillard, R. D. and Wimmer, F. L. (1978), *J. Chem. Soc., Chem. Commun.* 936.
1979   Shimba, S., Fujinami, S. and Shibata, M. (1979), *Chem. Letters*, 783.

Bibliography of the works of George B. Kauffman used in this book

1959 'Sophus Mads Jørgensen (1837–1914): A Chapter in Coordination Chemistry History', *J. Chem. Educ.* **36**, 521–527; reprinted in (1965), *Selected Readings in the History of Chemistry*, ed. by A. J. Ihde and W. F. Kieffer, pp. 185–191. Journal of Chemical Education, Easton, Pennsylvania.

1960 'Sophus Mads Jørgensen and the Werner–Jørgensen Controversy', *Chymia* **6**, 180–204.

1962 (with Beck, A.) 'Nikolai̯ Semenovich Kurnakov (1860–1941)', *J. Chem. Educ.* **39**, 44–49; reprinted in (1965), *Selected Readings in the History of Chemistry*, ed. by A. J. Ihde and W. F. Kieffer, pp. 191–196. Journal of Chemical Education, Easton, Pennsylvania.

1963 'Terpenes to Platinum: The Chemical Career of Lev Aleksandrovich Chugaev', *J. Chem. Educ.* **40**, 656–665.

1966a 'Foundation of Nitrogen Stereochemistry: Alfred Werner's Inaugural Dissertation', *J. Chem. Educ.* **43**, 155–165.

1966b 'Alfred Werner − Architect of Coordination Chemistry', *Chemistry* **39** (12), 14–18.

1966c *Alfred Werner − Founder of Coordination Chemistry*, Springer-Verlag, Berlin-Heidelberg-New York.

1967a *Werner Centennial*, ed. by G. B. Kauffman, American Chemical Society, Washington DC.

1967b 'Some Lesser Known Aspects of the Work and Thought of Alfred Werner', in *Werner Centennial*, ed. by G. B. Kauffman, pp. 41–69. American Chemical Society, Washington DC.

1967c 'Alfred Werner's Coordination Theory − A Brief Historical Introduction', *Educ. Chem.* **4** (1), 11–18.

1967d 'Alfred Werner's Habilitationsschrift', *Chymia* **12**, 183–187.

1967e 'Alfred Werner: Contributions to the Theory of Affinity and Valence', *Chymia* **12**, 189–216.

1968 *Classics in Coordination Chemistry, Part I: The Selected Papers of Alfred Werner*, Dover Publications, New York.

1970a 'Arturo Miolati (1869–1956)', *Isis* **61**, 241–253.

1970b   'Christian Wilhelm Blomstrand', in *Dictionary of Scientific Biography*, Vol. 2, ed. by C. C. Gillispie, pp. 199–200. Charles Scribner's Sons, New York.

1971   'Carl Ernst Claus', in *Dictionary of Scientific Biography*, Vol. 3, ed. by C. C. Gillispie, pp. 301–302. Charles Scribner's Sons, New York.

1972a   'The Stereochemistry of Trivalent Nitrogen Compounds: Alfred Werner and the Controversy over the Structure of Oximes', *Ambix* **19**, 129–144.

1972b   'An American Pioneer in Platinum Metal Research: The Life and Work of Wolcott Gibbs', *Platinum Met. Rev.* **16**, 101–104.

1972c   'Werner, Kekulé, and the Demise of the Doctrine of Constant Valency', *J. Chem. Educ.* **49**, 813–817.

1972d   'John Albert Newton Friend', in *Dictionary of Scientific Biography*, Vol. 5, ed. by C. C. Gillispie, pp. 189–190. Charles Scribner's Sons, New York.

1972e   'Thomas Graham', in *Dictionary of Scientific Biography*, Vol. 5, ed. by C. C. Gillispie, pp. 492–495. Charles Scribner's Sons, New York.

1972f   'Frederick Augustus Genth', in *Dictionary of Scientific Biography*, Vol. 5, ed. by C. C. Gillispie, pp. 349–350. Charles Scribner's Sons, New York.

1972g   *Resolution of the Tris(1,10-phenanthroline)nickel(II) Ion*, Willard Grant Press, Boston, Massachussetts.

1973a   ' "Quinquevalent" Nitrogen and the Structure of Ammonium Salts: Contributions of Alfred Werner and Others', *Isis* **63**, 78–95.

1973b   'Alfred Werner's Research on Polynuclear Coordination Compounds', *Coord. Chem. Rev.* **9**, 339–363.

1973c   'Alfred Werner's Research on Structural Isomerism', *Coord. Chem. Rev.* **11**, 161–188.

1973d   'Alfred Werner's Theory of Acids, Bases, and Hydrolysis', *Ambix* **20**, 53–66.

1973e   'Crystals as Molecular Compounds: Paul Pfeiffer's Application of Coordination Theory to Crystallography', *J. Chem. Educ.* **50**, 277–278.

1973f   'Paul Pfeiffer: Crystals as Molecular Compounds', *J. Chem. Educ.* **50**, 279–280.

1973g   'Heinrich Ley (1872–1938) and His Concept of Inner Complex Salts', *J. Chem. Educ.* **50**, 693–697.

1973h   'Heinrich Ley: On Inner Metal-Complex Salts. I', *J. Chem. Educ.* **50**, 698–700.

1973i   'Sophus Mads Jørgensen', in *Dictionary of Scientific Biography*, Vol. 7, ed. by C. C. Gillispie, pp. 179–180. Charles Scribner's Sons, New York.

1973j   'A Russian Pioneer in Platinum Metals Research: The Life and Work of Lev Aleksandrovich Chugaev', *Platinum Met. Rev.* **17**, 144–148.

1974a   (with Lindley, E. V., Jr), 'A Classic in Coordination Chemistry: A Resolution Experiment for the Inorganic Laboratory', *J. Chem. Educ.* **51**, 424–425.

1974b   'Alfred Werner's Research on Optically Active Coordination Compounds', *Coord. Chem. Rev.* **12**, 105–149.

1974c   'Paul Karrer — Rotating and Resolving: A Tragicomic Popular Play', *Chemistry* **47** (4), 8–12.

1974d   'Heinrich Gustav Magnus', in *Dictionary of Scientific Biography*, Vol. 9, ed. by C. C. Gillispie, pp. 18–19. Charles Scribner's Sons, New York.

1974e   'Early Theories of Metal-Ammines: A Brief Historical Review from Graham (1837) to Claus (1856)', *J. Chem. Educ.* **51**, 522–524.
1974f   'Alfred Werner's Views of Oxonium Salts', *Ambix* **21**, 227–246.
1974g   'Paul Pfeiffer', in *Dictionary of Scientific Biography*, Vol. 10, ed. by C. C. Gillispie, p. 578. Charles Scribner's Sons, New York.
1975a   'Alfred Werner's Research on Geometrically Isomeric Coordination Compounds', *Coord. Chem. Rev.* **15**, 1–92.
1975b   'Christian Wilhelm Blomstrand (1826–1897), Swedish Chemist and Mineralogist', *Ann. Sci.* **32**, 12–37.
1975c   'Frederick Augustus Genth and the Discovery of Cobalt-Ammines', *J. Chem. Educ.* **52**, 155–156.
1975d   'The First Resolution of a Coordination Compound', in *Van't Hoff–Le Bel Centennial*, ed. by O. B. Ramsay, pp. 126–142. American Chemical Society, Washington DC.
1975e   'The Discovery of Optically Active Coordination Compounds – A Milestone in Stereochemistry', *Isis* **65**, 38–62.
1975f   'Gustav Magnus and His Green Salt', *Platinum Met. Rev.* **20**, 21–24.
1975g   (with Myers, Robin D.), 'The Resolution of Racemic Acid: A Classic Stereochemical Experiment for the Undergraduate Laboratory', *J. Chem. Educ.* **52**, 777–781.
1976a   'Alfred Werner', in *Dictionary of Scientific Biography*, Vol. 14, ed. by C. C. Gillispie, pp. 264–272. Charles Scribner's Sons, New York.
1976b   *Classics in Coordination Chemistry, Part II: Selected Papers (1798–1899)*, Dover Publications, New York.
1976c   (with Lindley, E. V., Jr), 'The Resolution of *cis*-Bromoamminebis(ethylenediamine)cobalt(III) Bromide', *Inorg. Syn.* **16**, 93–97.
1976d   'Through the Back Door: The Role of Chance in the Choice of a Career of Two Coordination Chemists', *J. Chem. Educ.* **53**, 445.
1976e   'An Ingenious Impudence: Alfred Werner's Coordination Theory', *J. Chem. Educ.* **53**, 445–446.
1976f   'Alfred Werner's Research in Organic Stereochemistry', *Naturwissenschaften* **63**, 324–327.
1976g   'Il'ya Il'ich Chernyaev's Research on the Platinum Metals', *Platinum Met. Rev.* **20**, 126–130.
1976h   (with Golovnya, V. A., Leonova, T. N. and Craig, W.), 'Il'ya Il'ich Chernyaev (1893–1966): Some Recollections of His Work and Personality', *Ambix* **23**, 187–198.
1977a   'Il'ya Il'ich Chernyaev and the Trans Effect', *J. Chem. Educ.* **54**, 86–89.
1977b   'Left-handed and Right-handed Molecules: Louis Pasteur's Resolution of Racemic Acid', *Chemistry* **50** (3), 14–18.
1977c   'Coordination Chemistry: Its History through the Time of Werner' (audiotape and book), American Chemical Society, Washington DC.
1977d   'Christian Wilhelm Blomstrand (1826–1897) and Sophus Mads Jørgensen (1837–1914): Their Correspondence from 1870 to 1897', *Centaurus* **21**, 44–64.
1977e   'Early Experimental Studies of Cobalt–Ammines', *Isis* **68**, 392–403.

1978a   *Classics in Coordination Chemistry, Part III: Twentieth-Century Papers (1904–1935)*, Dover Publications, New York.

1978b   (with Benfey, O. T.), 'La Sintesis de la Urea por Wöhler – Una Retrospectiva en Su Sesquicentenario', *Rev. Chil. Educ. Quim.* 3, 284–291.

1979a   (with Benfey, O. T.), 'The Birthday of Organic Chemistry', *J. Coll. Sci. Teaching* 8, 148–151.

1979b   (with Chooljian, S. H.), 'The Artificial Synthesis of Urea: A Classic Experiment for the Undergraduate Laboratory', *J. Chem. Educ.* 56, 197–200.

1979c   'Alfred Werner's Early Views of Valence', *J. Chem. Educ.* 56, 496–499.

1979d   'Ferroceno, El Primer Compuesta Sandwich – Una Breve Revision', *Rev. Chil. Educ. Quim.* 4, 107–113.

1979e   'Las Ideas de Alfred Werner sobre Valencia', *Rev. Chil. Educ. Quim.* 4, 148–156.

Appendix E

Supplementary reading

Bailar, J. C., Jr, 'The Early Development of the Coordination Theory', in *The Chemistry of the Coordination Compounds*, J.C. Bailar Jr (ed.), Chapt. 2, Reinhold Publishing Corp., New York (1956). A short chemical-historical treatment of theories of complex compounds through the time of Werner (19 pp.).

Basolo, F., and Johnson, R., *Coordination Chemistry*, W. A. Benjamin, New York (1964). A short introduction intended as supplementary reading for general chemistry students (180 pp., paperback).

Graddon, D. P., *An Introduction to Co-ordination Chemistry*, Pergamon Press, New York (1961). A short but scholarly introduction to the subject (111 pp.).

Grinberg, A. A., *An Introduction to the Chemistry of Complex Compounds*, transl. by J. R. Leach; D. H. Busch and R. F. Trimble Jr (eds), Pergamon Press, New York (1962). A comprehensive treatment emphasizing history and experimental data, especially the work of Russian chemists (363 pp.).

Jones, M. M., *Elementary Coordination Chemistry*, Prentice-Hall, Englewood Cliffs, NJ, USA (1964). A comprehensive introduction for chemists and workers in other fields (473 pp.).

Kauffman, G. B., *Alfred Werner—Founder of Coordination Chemistry*, Springer-Verlag, Berlin-Heidelberg-New York (1966). A nontechnical biography (127 pp.).

Kauffman, G. B., *Classics in Coordination Chemistry, Part I: The Selected Papers of Alfred Werner*, Dover Publications, New York (1968). This book contains annotated translations of Werner's six most important articles, each with an introductory essay and commentary (190 pp., paperback).

Kauffman, G. B. (ed.), *Werner Centennial*, American Chemical Society, Washington DC (1967). A collection of forty-two papers, historical as well as technical, on coordination chemistry presented at the Werner Centennial Symposium, 152nd National Meeting of the American Chemical Society, New York, 12–16 Sept. 1966 (661 pp.).

Kauffman, G. B., *Classics in Coordination Chemistry, Part II: Selected Papers (1798–1899)*, Dover Publications, New York (1976). This volume contains annotated translations, each with an introductory essay and commentary, of many of the articles discussed in this book. The authors represented are: Tassaert, Magnus, Zeise, Graham, Claus, Blomstrand and Jørgensen (174 pp., paperback).

Kauffman, G. B., *Classics in Coordination Chemistry, Part III: Twentieth-Century Papers (1904–1935)*, Dover Publications, New York (1978). This volume contains annotated translations, each with an introductory essay and commentary, of many of the articles discussed in this book. The authors represented are: Ley, Chugaev, Friend, Pfeiffer, Wyckoff and Posnjak, Dickinson, Sidgwick, Chernyaev and Mills and Quibell (236 pp., paperback).

Lewis, J. and Wilkins, R. G., *Modern Coordination Chemistry: Principles and Methods*, Interscience Publishers, New York (1960). Six essays on thermodynamics, kinetics, isomerism, spectra and magnetochemistry (487 pp.).

Martin, D. F. and Martin, B.B., *Coordination Compounds*, McGraw-Hill, New York (1964). A short, elementary introduction at the beginning undergraduate level (99 pp.).

Murmann, R. K., *Inorganic Complex Compounds*, Reinhold Publishing Corp., New York (1964). A short monograph intended as supplementary reading for general chemistry students (120 pp., paperback).

Nathan, L. C., *A Laboratory Project in Modern Coordination Chemistry*, Brooks/Cole Publishing Co., Monterey, Calif., USA (1981). An inorganic chemistry laboratory manual for the advanced undergraduate (91 pp., paperback).

Nicholls, D., *Inorganic Complexes*, John Murray, London (1974). An introduction intended primarily for sixth-form students and teachers; includes test-tube-scale experiments and questions (76 pp., paperback).

Orgel, L. E., *An Introduction to Transition-Metal Chemistry: Ligand-Field Theory*, 2nd edn, John Wiley and Sons, New York (1966). A brief treatment of this important theory (186 pp.).

Quagliano, J. V. and Vallarino, L. M., *Coordination Chemistry*, D. C. Heath and Co., Lexington, Mass., USA (1969). A condensed treatment for general chemistry students. Includes exercises and problems (124 pp., paperback).

Read, J., *Humour and Humanism in Chemistry*, G. Bell and Sons, London, 1947, pp. 262–284. Read's personal reminiscences of his stay in Zürich as one of Werner's *Doktoranden*.

Subject Index

Name Index